THE FARMER and THE BUREAUCRATS

Barclays Bank dispossessed Watcyn Richards of his Pembrokeshire dairy farm and put him and his family on the road, but at least he was able to sue their solicitors successfully.

He found sacks of highly confidential documents of the affairs of the customers of the Nat West Bank blowing about in his wood long before the private financial affairs of Norman Lamont were made public.

He had stand-up battles with all sorts of authorities, from the Milk Marketing Board to the Ministry of Agriculture, and from the Health and Safety Executive to the Local Planning Authority, who lost their case against him at public enquiry and were ordered to pay the costs for their stupidity.

He did not win all his battles, but he won enough of them to come out with head unbowed, by no means broken, and most precious of all, with his family life intact and closely knit.

Roscoe Howells, himself a former farmer, is the author of fifteen books, and was for many years a columnist who wrote with scathing wit and trenchant humour. For the twenty years during which Watcyn Richards was waging his various campaigns against bumbledom and overweening bureaucracy, the author was privy to much of his business. He tells now of some of these confrontations in a book which will be read eagerly, not only by those who have had their own brushes with officialdom, but by all those who enjoy a good story told by a born raconteur.

Also by Roscoe Howells:

Non-fiction

Cliffs of Freedom
Farming in Wales
The Sounds Between
Across The Sounds
What Price Abortion? (autobiograhical)
Total Community
Old Saundersfoot
Tenby Old and New
Caldey
Farewell the Islands

Fiction

Heronsmill
Crickdam
Roseanna
Rhiannon

ROSCOE HOWELLS

THE FARMER
and
THE BUREAUCRATS

EMISSARY PUBLISHING
P.O. Box 33, Murdock Road,
Bicester, Oxon., OX6 7PP, England.

First published in Great Britain September 1993 by
Emissary Publishing,
P.O. Box 33, Murdock Road,
Bicester, Oxfordshire, OX6 7PP, England.

British Library Cataloguing-in-Publication Data.
A catalogue record for this book is available from the British Library.

ISBN 1-874490-11-2

© Roscoe Howells 1993

All rights reserved.

Roscoe Howells asserts his moral right to be identified as the Author of this work in accordance with the Copyright, Designs and Patents Act, 1988.

This book may not be reproduced, in whole or in part, in any form (except by reviewers for the public press), without the prior permission in writing of the publishers.

Printed in Great Britain by
Manuscript ReSearch, Oxfordshire, England.

For
RAY
Whose loyalty and strength made survival possible

CONTENTS

Page

PREFACE:	*Bank Documents That Wandered*	9
Chapter One:	*How it all started*	15
Chapter Two:	*The road that led to London*	30
Chapter Three:	*Cover-up and apathy*	45
Chapter Four:	*A difference of opinion*	53
Chapter Five:	*Some have to stand alone*	66
Chapter Six:	*The affair of the tractor safety cab*	81
Chapter Seven:	*Tests don't come easy*	98
Chapter Eight:	*What's it worth?*	115
Chapter Nine:	*Rising cost and papers found*	123
Chapter Ten:	*Not a satisfactory settlement*	145
Chapter Eleven:	*A most unusual planning matter*	162
Chapter Twelve:	*A shot-gun licence and more pollution*	168
Chapter Thirteen:	*A little white boy in the coal shed*	174
Chapter Fourteen:	*Thou shalt not fart in Church*	186
Chapter Fifteen:	*How high do you bounce?*	192

PREFACE
Bank documents that wandered

On the first of February, 1987, the Sunday Express carried a story under the headline, **'Farmer digs in over lost bank papers.'** It was what is known in the newspaper world as a scoop and it was only a matter of hours before the farmer was being descended upon by journalists and photographers from local papers, radio and television, like flies round a honey pot.

The report said:

A farmer who stumbled across confidential bank documents on his land is demanding a £50,000 'reward' for their return.

Watcyn Richards has buried the papers on his property as the National Westminster bank prepares itself for action.

The documents include details of the financial affairs of a judge.

Mr Richards, of Camrose, near Haverfordwest, Dyfed, discovered the haul on the way home from a night fishing expedition.

A piece of paper blew against his chest, and shining the torch on it he saw it bore the trading figures of a local business man.

Next morning he went back to the spot – part of his 100 acre farm which is being infilled with builder's rubble. Amongst the broken bricks were several split black plastic sacks, with documents scattered about.

When Mr Richards examined their contents he found: Blank credit cards.

Bundles of bank credit statements, some bearing the names of well-known people in the area. One revealing a business man's £9 million turnover.

Detailed notes of personal interviews.

Documents showing a local farmer had borrowed £300,000 to buy another farm.

So, Mr Richards, 47, photocopied part of his find and, through a friend, contacted the branch at Haverfordwest.

<u>*Bags*</u>

Next morning the manager, Mr John Jones, plus an assistant arrived at the house where the farmer lives with his wife and three children.

'He went white when he saw the documents,' said Mr. Richards.

The manager explained the sacks had inadvertently been taken during renovations to the bank.

But at a later meeting Mr Richards claims he was offered 'Out of pocket expenses' for returning them.

So he put most of the important papers into water-proof fertiliser bags and buried them on the farm.

Now National Westminster's solicitors in London have written to the farmer giving him seven days to return the goods.

They point out that the papers legally belong to them.

But Mr Richards is not kindly disposed to banks after a problem £180,000 overdraft.

'I intend to have a reward,' he said.

'I don't want blood, I want enough so I can save my farm and be put in a reasonable trading situation.'

A spokesman for National Westminster said:
'Some confidential paper waste was accidentally removed with other waste awaiting disposal from the bank premises at Haverfordwest.'

The report, picked up over the telephone, was a model of factual reporting, and I would draw attention to two slight inaccuracies only. Watcyn had not been fishing but rabbiting.

And we do not, nor ever did, speak of Dyfed down here in Pembrokeshire.

The bit about Mr Richards not being kindly disposed to banks had to be in the running for the prize for the understatement of the year. Reference was made in the report to his problems with a £180,000 overdraft. It was already generally known in the area that Watcyn was one of a distressingly increasing number of the farming fraternity in West Wales who were having to jump through Barclays' financial hoop.

Year in and year out, for something like forty years after the war, farmers had been invited to attend conferences, discussions, field days, farm walks, seminars and meetings where the gospel had been expounded that bank borrowing was the cheapest form of borrowing, and that the bank manager was the farmer's best friend. The only trouble was that even his best friend wouldn't tell him. Time was when people believed this.

Nowadays it is by no means unusual to meet up with respectable members of society who will tell you that banks are rogues and robbers.

It is an old-fashioned notion indeed that banks are there to advise or serve the customer. They are there to make money. And if they pour too much down the drain in bad judgement abroad, or in being fooled by the Robert Maxwells of life, then they have to bludgeon it out of those smaller fry at home who are unable to hit back.

Back in the early 1970's a bank manager who was a good personal friend of mine told me that he had just come back from Swansea where he had been attending a regional meeting of bank managers. And the atmosphere had been one of gloom and disillusionment. Men, he said, who had for years been happy in their jobs, believing they were doing something worthwhile, had become sickened at what was happening to banking and, at fifty years of age, were looking for early retirement.

At a crucial stage in Watcyn's battle with Barclays, they had told him that they had lost all his papers and records.

How was he to know that they were not blowing about on somebody else's rubbish tip and that Barclays were lying anyway?

It was the misfortune of the Nat West manager in this particular case that his bank's documents had fallen into the hands of a man in Watcyn's position who had every reason to hate the very term bank manager.

The manager, clearly a very worried man, knowing only of Watcyn's differences with Barclays, told him to come in and discuss the position and they would sort everything out for him. And this was said in front of a highly credible witness.

A drowning man will clutch at any passing straw but, when Watcyn went into the bank, the manager had changed his tune.

There is an old Pembrokeshire saying that a dying pig will kick. So Watcyn went to the Press. Whether he was wise to do so at that stage is open to question, but the business had been going on for more than a month and it does explain in some measure the talk of a reward being demanded.

For about fifteen years before all this happened Watcyn had suffered more than his share of trials and tribulations, and I believe that the story of some of them is worth recording. 'To know all makes one tolerant.'

His troubles had not been only with the bank. There was far more to it than all that. He had perhaps had more than his fair share of the frustrations and stupidities which come to us all from time to time at the hands of bureaucracy. We are fortunate indeed if we can avoid them altogether. Most of us, I suspect, tend to take the line of least resistance and, in the end, however worthy we may be, sigh with frustration and give it up as a bad job. At the best we are likely, if not to walk away from an apparent brick wall, then to try to go round it. Watcyn Richards has a somewhat different philosophy. His approach to the obstacle of a brick wall would appear to be to put his head down and go through it.

Nice Mr Major has now decreed that these bureau-cratically-created difficulties should be swept away. Not before time some would say. Watcyn could no doubt tell him of one or two places where it would be possible to make a useful start.

CHAPTER ONE
How it all started

According to Messrs Flanders and Swann, it was on a Monday morning the gasman came to call. And it presaged a whole saga of misfortune.

In my case it was on a Sunday evening that Watcyn Richards came to call and, nearly twenty years later, rather like the saga of the gasman, there would seem to be no end to it.

It began, as I recall, after chapel on the evening of Sunday, November 4th, 1973.

I suppose there are people in life who are accident prone or who attract troubles and misfortune more so than others, and possibly Watcyn is one of them. On the other hand, there are those who do not suffer fools gladly, tend to speak their minds when they come up against graft or incompetence, and do not compromise and walk away in preference to making a nuisance of themselves. Watcyn would certainly come into that category.

He had been given my name by a mutual friend, Derek Rees, a journalist with the Evening Post, and had telephoned me earlier in the day with a tale of woe I found hard to believe.

Derek, who had already written a story about Watcyn's troubles under the Ministry of Agriculture's brucellosis scheme, knew that I had learned overmuch about brucellosis the hard way.

For the uninitiated, brucellosis is the euphemism by which contagious abortion in cattle is known, and it can also affect humans. Earlier that year my book, *What Price Abortion?*, had been published. I had seen my life's work as a Guernsey breeder in ruins, my health had been shattered when I had contracted the disease myself, and it was known that I was not greatly enamoured of the Min. of Ag. and Fish. and the blinkered bureaucratic stupidity of some of its officials.

Even so, I had no intention of rushing in with offers to help. I wanted to meet the man first. So he came to see me that Sunday evening and was waiting for us when my wife and I came home from chapel. There are some people we take to on sight. In my case, Watcyn was one of them.

A big, strong-looking type, an inch or so over six feet, with dark, Celtic features and a warm smile, his black hair has gone grey, indeed almost white, since that first meeting, and he has experienced enough trouble for that to be unsurprising.

I sensed straightaway that he was genuine, and we have been good friends ever since.

Yes, he does tend to get carried away sometimes, and he doesn't always stick to the point in an argument, but I have always found him to be honest.

Brought up, from the age of six, by the sea-side at Solva, in the north-west of Pembrokeshire, he spent his childhood climbing the rocks, exploring the pools and the water's edge, and becoming imbued with a feel for the countryside and the creatures of the wild. Then, on leaving school, he learned his trade as a butcher, before going to work on a farm.

Eventually he worked as a cowman for a large-scale dairy farmer whose methods, before he packed up and went to Australia, were a byword in the county, so at least Watcyn could claim that he knew how the job should not be done.

In 1969 he acquired the tenancy of Bunkers Hill, an eighty acre farm near Camrose in Pembrokeshire, on the banks of the Western Cleddau, and with land also adjoining the Old Mill Stream which runs down from Leweston. There were otters in the river, badger setts in the woods, and he was well content with his portion here below. He was twenty-nine years of age at the time, with a wife and a son of five. Two daughters were to be born later.

In 1972 he bought a twenty acre holding without buildings, known as White Thorn, but more particular mention will be made of this later.

At that time the Ministry of Agriculture were experiencing the first, fine careless rapture in the launching of their Brucellosis Accredited Herds Scheme. Watcyn joined enthusiastically and his was only the sixth herd in the county to become accredited. So, at that stage, and in that respect, he would not have been frowned upon by the Ministry, however much they may have had cause to cease to love him since.

Trouble was not long in coming. It began in the summer of 1972 when he was producing over 120 gallons of milk a day.

Within the year it was down to 40 gallons a day, eight animals had died, some had aborted, and 28 valuable dairy cows had been slaughtered. There had been a complete breakdown in the brucella test, and the Ministry officials had placed an order on him forbidding him to use twenty-three acres of his best grazing land alongside the river. Easy for them, of course.

Six weeks holiday, five day week, index-linked pension, keep at it long enough and you'll probably finish up in the Honours Lists.

Why had they stopped him from using the twenty-three acres of grazing land? Well, of course, because of the pollution, but that was nothing to do with the Ministry because the river was the sphere of the River Authority. Quote: 'We sympathise with Mr Richards who is in a very serious position, but we can do nothing for him.'

The Public Health Inspector of the Rural District Council was no more helpful, but confirmed what the Ministry man had said. Yes, of course, very much a matter for the River Authority but, most definitely, as a public health authority, they would be interested if there was a public health hazard.

The Min. of Ag. and Fish., however, did make one special concession. They issued him with a licence to use the live vaccine S19 on his adult stock. Had I known him at the time I could have assured him that it would be as much use as putting red pepper on a gumboil.

With experience he was to find that the River Authority were just about as useful and helpful. They could equip their bailiffs with expensive binoculars and cameras which would operate in the dark to try to catch the occasional poacher, but try to agitate them into doing something when thousands of fish came floating down the river, bellies upwards, when it was a case of pollution, and it was a different story altogether.

In 1980, by which time he had become considerably wiser, Watcyn drew the attention of the River Authority to a case of pollution where farm and silage effluent were being piped directly into the County Council's drainage system which discharged into a stream flowing into the Western Cleddau. The divisional solicitor to the River

Authority saw the evidence of it for himself and said that definitely something would be done about it.

The evidence he saw was some County Council roadmen emerging from their lorry and there, before his very eyes, as the saying goes, lifting a couple of manhole covers and clearing out the muck and slurry which were blocking the drains. This had been causing the slurry to run down the road, and people had started to complain. So the County Council, as the Highway Authority, sent their stalwarts along to see to affairs. And they did. Before the very eyes of the divisional solicitor to the River Authority. He said he couldn't believe what he was seeing, and that was why and when he said that definitely something would be done about it. Watcyn said he wouldn't dare and bet him a couple of pints that, if he did dare to try, he would be moved from the district. He must have tried, because he was moved from the district shortly afterwards, nothing was done, and Watcyn didn't even have his two pints. And in case anyone is looking for a transfer, the same system is still in operation.

The pollution which had been the cause of Watcyn's horrendous troubles was coming from a neighbouring farm where a slurry lagoon had been sited, with Ministry approval, far too close to a watercourse. It is always sad when this sort of thing happens between neighbours, but even sadder when a young man, trying to make his way in the world, suffers such appalling losses as Watcyn experienced.

The officials of the River Authority, not being as enthusiastic as would have been implied by the Ministry officials and the Public Health Inspector, and seeming to be reluctant even to take samples from the Old Mill Stream, which flows into the Western Cleddau, Watcyn had to call in the public analyst himself. The report, dated July 12th, 1973, would have made the Devil's Schedule look like the Book of Common Prayer.

The Old Mill Stream was visited on a clear, sunny day and during a dry spell.

The stream was followed from its entry into the Western Cleddau for several hundred yards upstream.

Even at its confluence the stream contained much greyish white floating matter and it was malodorous. Conditions became progressively worse upstream until we came to a drainage ditch on the left hand side. Above this point of entry of the ditch the stream was entirely different in character being practically clear and odourless with the gravelly bottom brown in colour.

For at least one hundred yards below this source of pollution the stream was particularly choked with greyish white matter and the bottom was black in colour and evil smelling.

Boulders and stones were coated in thick deposits of the greyish white matter.

From a visual point of view the Old Mill Stream was suffering badly from polluting material entering from the drainage ditch referred to. At this time the inflow from this ditch was continuous but fairly small, but clearly much further polluting material was there which would be washed into the Old Mill Stream by the first rain.

The analytical findings showed the following:—

1. The Old Mill Stream above the point of pollution referred to was of fair quality. The organic matter which it contained had been oxidised to a significant extent and that which remained was not excessive and insufficient to cause a nuisance in this well oxygenated, cascading stream.

2. The Old Mill Stream below the drainage

ditch had deteriorated badly and now its organic content in solution was so high as to place it well beyond the worst category of normal classification into which streams and rivers were placed by a Royal Commission.

The stream could not cope with this large quantity of polluting matter and so its character had changed and now it bore a heavy growth of greyish white fungal filaments commonly referred to as sewage fungus.

3. The inflow from the drainage ditch contained so much organic matter (which in this case was the offending pollutant) that it was comparable at this time to a very raw domestic sewage.

In our opinion the Old Mill Stream was being so polluted by the flow from the drainage ditch that it had become unfit for use for watering for farm stock.

And still the River Authority did not want know. At that time it was the South West Wales River Authority. Later it was the Welsh Water Authority. The same people, of course, but more of them. Something about Parkinson's Law, I think.

As far as his losses were concerned, Watcyn had two detailed reports prepared by Basil Jones & Sons, the old established Haverfordwest firm of auctioneers, valuers and land agents. The first report covered losses which were sustained as a result of his not being allowed to graze certain fields, the extra labour in moving milking–cows twice daily to and from other pasture, the loss of hay and silage in the summer, thereby depleting the stocks of winter fodder, the loss of accredited herd status, and loss of some of the younger stock.

This valuation came to £801.70.

The second report was far more serious, covering as it did substantial losses of stock through death, losses involved through having to sell reactors to the butcher and, of course, loss of milk. Altogether this came to £5,819.23.

At that stage Watcyn's solicitor told him that, in order to mount a case against the neighbour, it would be necessary to engage a barrister. That costs the sort of money which young men who have just started farming and are overtaken by a calamity of this magnitude do not have. It was no more encouraging for Watcyn when his solicitor told him he could not possibly continue to face such losses and that he might as well pack up farming. At that point he lost his temper and called on his neighbour. What dire threats he made I do not know, but I have no doubt whatsoever that they would have been highly colourful and, to a certain extent, appeared to have been successful because a new lagoon was built. Unfortunately, this, too, was badly sited.

I mentioned earlier that the lagoon had been built with Ministry approval. In effect this means with the approval of the minion who comes to advise and to inspect the job. In this case, the one who agreed to the new lagoon was the one who had approved of the siting of the first effort.

But never mind. In due season he was able to shout the odds as a member of the local Farming and Wildlife Advisory Group, appeared in the Honours List and settled down to enjoyment of his index-linked pension.

A lovely old hymn it is in Sankey's book, and all together now, 'Tell me the old, old story . . . .'

Next we come to a complaint to the Member for Parliament (the then Hon Nicholas Edwards). He takes it up with the Secretary of State (the then Hon David Gibson Watt), who takes it up with the River Authority. They make their bureaucratic, fudging, sweep-it-under-the-carpet reply, and the Secretary of State writes to the local M.P. saying, 'I understand that there has been a considerable improvement etcetera and so on and so forth,' and the M.P. then writes to his constituent saying, 'I hope that the information is correct and that there has been a considerable improvement, and that you will be able to restock in the near future.' The only respectable comment in reply to which is 'Cobblers'. Or is that not respectable? It's a job to know these days with what passes on television but is not acceptable in polite society. That was in the September of 1973.

This, then, was the position when Watcyn came to see me in the November. The neighbour was now circumventing any possible trouble from the River Authority by syphoning the slurry out of the lagoon at week-ends. Officials of River Authorities do not normally work at week-ends as far as I know. Probably something to do with N.A.L.G.O.

In my case he had a sympathetic listener, and thereby hangs a tale.

A year or two previously I had been visited by a whatever-it-was-officer of the Carmarthen District Council, in which area we then lived, and he was looking for a spot of co-operation from the general public. I was the general public and he was a former police officer. In his new capacity, it seemed, he was determined to do something more useful than picking on the hapless and relatively law-abiding motorist. He was anxious to track down and come to grips with some of the nefarious characters who dump clapped-out refrigerators and electric cookers, along with old mattresses and other rubbish, in any lay-by or roadside verge which appeals to their passing fancy. Even if I, of the general public, could not ascertain their names, it would suffice if I could but obtain the number of their vehicle.

Well yes, we who dwell in the countryside are sick and tired of this loutish practice, but to identify the malefactors is more easily said than done. I sometimes think that the dumping must be done for them by the fairies or the little green men. Nobody ever seems to see anybody doing it. But it was mighty encouraging to know that somebody, somewhere, from the teeming ranks of officialdom was on our side.

Thereafter all returned to normal and its idyllic calm. Then one lovely spring day, down by the Cherry tree lay-by, near where we then farmed – and it is known as the Cherry Tree lay-by because a cherry tree used to grow there – I came upon a good neighbour of mine standing with a puzzled expression on his face and gazing at the latest offering. A sparkling brook babbles through the sylvan glade, and now, to my neighbour's bemusement, the delightful setting had been enhanced by a load of junk dumped from a garage which had obviously been having a massive clear-out. Cogs, Hardy-spicer joints, broken spanners, rusty lamps, fuses, big ends, little ends, crank shafts, oil-drums and tins various cascaded from the grass verge to the stream below.

My neighbour and I conferred. True, we had not witnessed the malefactors perpetrate this outrage, but this was one of those occasions when that did not really matter, because there were all manner of labels with the address of the garage on them.

So I, on behalf of the general public, went home and set the wheels of justice and retribution into their slowly grinding but inexorable motion. Or so I thought. The ex-policeman was certainly prompt to respond to my telephone call and came and collected a fistful of labels in no time at all before dashing off looking as happy as a lamb with two mothers. I had thought to write as happy as a dog with two lamp-posts, but decided eventually that it would not have been an apposite analogy because it could have led to the eventual impression that he was like a dog with a bone, which was something he certainly was not. He seemed to let go very quickly.

The date was March 20th. Perhaps I was a little impatient and not sufficiently understanding of how these matters have to be handled but, week after week, I scanned the columns of the local papers in vain. Never was there a mention of that in which I, as a member of the general public, now had more than a passing interest. So, on July 5th, I invested in another telephone call.

Yes, oh yes, he had indeed interviewed and taken a statement from the garage proprietor who had been most forthcoming and had not denied that all the junk had been his. However, he disclaimed any responsibility because he had paid money to some other outfit to take the stuff away and dispose of it.

Ah, well, what did these other characters have to say about it?

Yes, well, it was not as simple as all that because there were oil drums there, and the oil had washed into the stream, thereby causing pollution, and so it would have to be a joint prosecution in conjunction with the River Authority and he had passed it on to them. And what were they doing about it?

He had to admit that he hadn't heard anything further from them and he hadn't followed it up. You see what I mean about a dog with a bone?

Not being satisfied, I telephoned the River Authority myself. They were at Llanelli then, but I don't know where they are now. Not that it did me any good, because they knew nothing about it. Well, they wouldn't, would they? Not up there at Head Office. But they promised to send for the file and ring me back. Which they did. And it was to tell me that the complaint had indeed been received from Carmarthen R.D.C. and they had indeed sent their local officer to investigate. Apparently he had called at a cottage down the road to make enquiries but could not find anybody who knew anything about it and so there was nothing they could do. So I asked them could I now write a piece to say that people could now pollute the rivers with impunity as well as with waste oil and, judging by the way the telephone crackled, the chap at the other end must have thought I sounded like a low-down type who would rattle a packet of crisps at the opera. The only satisfaction I had was to tell him what I thought of the idiot they had sent out on such an enquiry and express the hope that I would never meet him. Such is the age in which we live that not only can parasites like this not be sacked, but are even, in due season, promoted on their eventual blissful way to drawing their index-linked pensions. The whole episode left me feeling frustrated and infuriated.

    It was exactly four months later that Watcyn came to see me and told me his story much as I have related so far. Then it transpired that the local man who had failed to take any effective action was the same pathetic character who had drawn his salary and mileage allowance for failing to do anything in my own case earlier in the year. Which is why I have said that Watcyn found in me a sympathetic listener.

    I knew that if he hoped to sort out his pollution troubles he would have to go much higher than this idiot. If he could not cope with a small case like ours there was never a chance that he would come to grips with a problem of the magnitude which now faced Watcyn, and I promised to do what I could to help.

    The following day I telephoned the Clerk to the River

Authority and had a bit of sense at last, together with a promise of action if the need should arise. Which it did, on the following Saturday night.

Watcyn telephoned to say that, true to week-end form, the pipe was back in the neighbour's lagoon syphoning the slurry into the ditch leading into the Old Mill Stream. It was about 10.30 p.m. when he telephoned and I told him to ring the River Authority's local man, tell him what was happening, and demand his presence and a spot of action forthwith. Ten minutes later he rang again to say that he had spoken to the little chap and he had refused point blank to stir from his house. So I told Watcyn to ring him once more and tell him that if he was not on his way in five minutes then I would ring my namesake at his home to put him in the picture and ask him what he was going to do about it. So maybe at this point I should explain about my namesake with whom I have often been confused.

Dr Roscoe Howells was one of the head-bosses with the River Authority. We are not related as far as I know. I am assured by those who know him that he is a very nice man, and I had reason at one time to believe that I was perhaps better looking than he was. So maybe I should explain about that as well, for up to the time of writing I have never met him.

Some years ago he made a statement concerning the incidence of salmon disease in Welsh rivers. Our national paper, *Two Minutes Silence*, carried the statement in full and must have thought it was rather an important statement at that because they also carried a picture of the handsome Dr Roscoe Howells. Unfortunately they used a picture of me. I didn't mind in the least, but I was none too sure how my namesake would feel about it. However, nothing was said, and he made no disclaimer, so I assumed he was satisfied and that was why I thought what I did about possibly being better looking.

Notwithstanding all of this, the name seemed to have worked the oracle because, in spite of any agreement with N.A.L.G.O. or anybody else, the little chap jumped to it. At Watcyn's insistence he took a sample of the water but would not call upon Watcyn's neighbour at that hour of midnight. With great courage he removed the pipe. By the following night it was once again in place syphoning the slurry into the ditch.

On the Monday morning I telephoned my namesake and spoke to him for the only time in my life.

These things take time but, by February, Watcyn's neighbour was in court and pleaded guilty. A cock-and-bull yarn went with it so that he got away with a conditional discharge, but it meant that Watcyn had sorted out that side of the problem.

It still left him, however, with the question of the brucellosis breakdown, which was why he had come to see me in the first place, and here, because of my own bitter experience, I was able to be of some help to him.

I said all I want to say on that subject in my book, 'What Price Abortion?', and would seek now to add only one comment. For some years, speaking from bitter first-hand experience, I had been a vitriolic critic of the Ministry of Agriculture's useless S19 vaccine. I was on record as having said that we would never clear brucellosis from our herds unless and until S19 was outlawed, as it had been in Switzerland and elsewhere. It was a live vaccine and, therefore, only blanketed the disease down, or, to put it another way, kept it in being. And I was proved right. To be proved right when you are in a minority of one is not the shortest and quickest route to popularity. But it is remarkable what comments have been made to me in the years since then. I would quote only one of them.

I cannot remember the occasion or the character concerned except that he was one of the older officials of the Min. of Ag. and Fish and Food. It was no longer of any great interest to me, but he told me that the Minister of Agriculture at the time of the original importation of S19 vaccine from America had a financial interest in the firm producing it and that they were none too stable. The edict went through the land, however, that the Ministry stipulated and promoted the use of S19 in this country and from then on the American firm never looked back.

Because of this experience I was able to tell Watcyn in November 1973 that he would have nothing but trouble unless he stopped using S19 and started to use M45/20, a dead vaccine being produced by Phillips Duphar. M45/20 had been one of the first vaccines used but, originally, it had been live and caused trouble. Now it had been killed, but the bureaucrats of the Ministry would have nothing to do with it. My own experience with, and knowledge of, M45/20 had come too late to be of any practical help to me, but it was of great help to Watcyn.

The Ministry told him that on no account would they authorise the use of M45/20 in his herd, so he went ahead and used it. It saved him from financial ruin, brought his trouble under control, and he came back into the Accredited Herds Scheme.

Maybe success on both scores gave him an unduly elevated idea of my ability to help. Be that as it may, it certainly meant that it was not the last I was destined to see of him.

CHAPTER TWO
The road that led to London

For a while everything went very quiet. Had I known then what I know now I would have realised that it was too good to last.

In 1975 Watcyn ran into trouble with antibiotics in his milk. It was the sort of thing that can happen to the best of us, and Watcyn took it seriously. His own family drank the milk he produced and he understood how serious and lethal it can be when milk is contaminated. He is, in fact, far more concerned about the dangers of antibiotics than many farmers I know.

It would be possible to write a long chapter on what has happened to the nation's health as a result of the abuse, over-use and misuse of antibiotics. There has been an element of irresponsibility in the farming community the same as elsewhere. There has also been the pathetically self-righteous element who have been fortunate enough to avoid trouble themselves, and who hold up their hands in horror and condemn everybody who has come up with a failed sample of milk. As often as not they are the staunch committee men who go through life convinced that the National Farmers' Union and the Milk Marketing Board can do no wrong.

Who was it said, *'Power tends to corrupt and absolute power corrupts absolutely?'*

Let there be no mistake about it. The advent of the Milk Marketing Board in the 1930's meant salvation for a virtually bankrupt farming industry, and they have done some sterling work since that time. That, however, was more than half-a-century ago, and we fool ourselves if we fail to recognise that since then they have acquired a whole heap of dead-wood at every level.

So we come to the question of the tests. In the old days there was the apocryphal story of the farmer who was charged with putting water in his milk and, when the prosecution pointed out that a trout had been found in the churn, the farmer pleaded that it was only circumstantial evidence.

Let us not carry our pleading in the present case as far as that. But one farmer in Watcyn's area, at the same time as Watcyn ran into trouble, had a notification from the M.M.B. to the effect that his solids and his butterfat were both below standard and, unless he did something about it, consideration would have to be given to the revocation of his licence. That could have been a serious matter had it not been for the fact that the farmer had given up producing milk five years previously. An attempt was made later to explain that it had been a mistake and that the notice should have been sent to one of his brothers also farming in the area, but, not only did they have his name and his address right, they had his old licence number right as well.

Since then examples have been legion of people who have been accused, tried and found guilty, even when no animals in the herd have been under treatment. In cases, however, where farmers have had notification of antibiotics in their milk when they are no longer producing milk, then even the M.M.B. have not been able to make out much of a case.

How, then, could such things have come about? As the result of quite a bit of fuss from the Watcyns of life, the M.M.B., a couple of years after Watcyn's trouble, changed their system of identifying samples. On April 1st 1978, and I hardly think there can have been any significance in the date, they introduced a system of issuing producers with labels. Each producer had his own labels, with his name and address and M.M.B. reference number, and these were available for the milk lorry driver to stick on the container with the milk sample.

The screed announcing the introduction of the system said, 'It is hoped that the introduction of this universal system of labelling of samples will not only assist the tanker driver, but will give you every confidence in the identification of your samples.'

Prior to that the system had been for the driver to write with a pencil on the rough surface of the bottle containing the sample. And it was not unusual for a driver to pull up at a farm and, having filled a few bottles with samples, write on the names of the previous farms where he had called and forgotten to take a sample.

Unaware in the initial stages of some of these factors, Watcyn took matters seriously and was a worried man. He just could not see where he was going wrong. As recommended in the notes which came with the notification he called in the M.M.B.'s local whiz-kids. The fact that they were useless took some time to establish and was of no help to Watcyn, either before or after he had realised it. He had carried out faithfully the manufacturers' instructions for the various drugs he had used and was therefore bewildered, just as the whiz-kids were. The crucial difference was that they went on drawing their salaries whilst Watcyn was having heavy penalty deductions from his milk cheque.

The one whiz-kid, who wore a long, coloured scarf, 'like a college scarf,' as Watcyn put it, did not endear himself to Watcyn when he went off to Haverfordwest to the dairy and returned with some literature pertaining to an entirely different antibiotic from the one which Watcyn was using. Watcyn made some apt comment on the man's stupidity, an altercation followed, and was concluded when Watcyn threatened to hang him by his scarf from the nearest so-and-so tree. It was some time before the scarf even communicated with Watcyn again, and that will come a little later.

The first failure came on Dec 1st, 1974, when a cow was being treated with Oxycomplex for foul in the foot. In April there was another failure for the same reason and Watcyn asked his vet for a new prescription, which resulted in a change to Terramycin.

In October, using Terramycin, again for foul in the foot, there was another failure and Watcyn contacted the M.M.B. In response to his call for help the scarf arrived. He took a test which proved negative, even though, by way of an experiment, the cow had been treated and the milk not withheld. The scarf was bemused and it gave rise to a totally unfounded suggestion later that Watcyn was refusing to withhold milk from cows under treatment.

In January of 1976 there was a further failure, but this was for a different reason. This time it was mastitis and there had been an intra-mammary injection of penicillin. In ignorance, Watcyn had been using a quarter-milker. This device had been developed initially at the behest of the M.M.B. to enable the infected quarter to be milked separately and the milk from that quarter to be withheld. Subsequently it was discovered that it was necessary to withhold all the milk from a cow being treated, but we cannot blame the ignorant farmer for not knowing about this until it was found out the hard way.

In March a cow was under treatment for mastitis and there was another failure. Then there was a further failure in April, but no cow was being treated. This failure was unexplained, but the facts cut no ice with the inquisitors, and, with such a record, Watcyn was already for the high jump.

Amongst the notes sent out by the M.M.B. in an attempt to save the producer's soul from perdition there was one which read:

Intra–mammary treatment should be carried out in accordance with veterinary advice or in accordance with the instructions of the manufacturers of the preparation, both in relation to the method of treatment and to the withholding of milk for sale for the appropriate period. The minimum withholding time should be given in the instructions issued with the preparation, and the milk which is withheld may be fed to stock.

Please note the bit about feeding the contaminated milk to stock because it will be germane to an episode which occurred later.

We must confine ourselves for the moment to what was happening about the tests various. There is an old Pembrokeshire story about a simple–minded youth who slept with his father and mother until he was twenty–one. On his twenty–first birthday his father suggested to the young man's mother that they should move the son into another bed as he thought the boy was beginning to take notice. Watcyn was certainly beginning to take notice.

What he noticed, amongst other interesting points, was that on the data sheet referring to Terramycin Q–100 Injectable Solution, put out by Pfizer Ltd, it said that 'milk drawn from treated cows during treatment or for a period of 60 hours after treatment should not be used in the manufacture of cheese.' And Watcyn's perfectly reasonable question in the face of that was how should he be expected to know whether his milk was going for cheese or not? The withholding period, it seemed, for milk going to the liquid market, was 48 hours.

There was something else that Watcyn noticed, too. The withholding time, as advised by the manufacturers, was based on a standard antibiotic level of 0.05 iu/ml milk. This recommendation was made by the Veterinary Products Committee, (two thirds of whose members are connected with pharmaceutical companies!), accepted by the Min. of Ag. and Fish. and Food, and rubber-stamped. Just like that. Pfizer's Trial Data sheet explained that *'a trial was carried out to determine the oxytetracycline hydrochloride content of milk subsequent to the administration of Terramycin Q-100 Injectable Solution. Twelve Friesian cows, each in mid-lactation were allotted to four treatment groups with three cows per treatment.'* Very comprehensive. And quite enough to satisfy the Min. of Ag. and Fish. and Food.

So, with a comprehensive test like that, maybe Watcyn could be forgiven for thinking that if he followed the manufacturers' instructions, and withheld the milk from a treated cow for the time stated on the bottle containing the drug, then everything in the garden would be lovely and Bob's your uncle. Dozens and hundreds of poor fools in the farming community have gone through life blissfully imagining the same thing.

Then, however, continuing to take notice, Watcyn realised something else. The M.M.B. standard was set at 0.02 iu/ml, which was a more stringent standard altogether.

And it is a sad, but harsh, fact of life that, at this standard, milk should be retained for much longer than the period recommended by the manufacturers. In fact, a sample could fail more than a week after a cow has been treated.

It was whilst Watcyn was puzzling over some of these anomalies that he was summoned to appear before his betters and that was where he again came into my life. I said earlier that the pollution business had left him with an unduly elevated idea of my ability to help.

He was summoned to appear before the Hearings Committee of the M.M.B. to give reasons, if any, why his licence to sell milk should not be withdrawn. And the hearing was to take place at Thames Ditton. Look at the map and consult the time-tables. Watcyn was also in the middle of making silage.

He wrote to say that he would like to appear before the committee, but that he was in the middle of making silage, was self employed with nobody working for him, it would take time to find somebody who could milk for him during his absence, and that it would be almost impossible to arrange to travel to Thames Ditton at such short notice.

There came a reply, very civil, from the Secretary of the Board, no less, saying that the hearing had now been changed from May 6th to June 16th and the venue had been changed from Thames Ditton to the Russell Hotel. A subsequent letter explained, 'It may be helpful to note that the Russell Hotel is very close to Russell Square underground station.' Just how helpful it was, the Secretary could never have imagined.

It may be thought that, for the convenience of the farmer, the hearing should be held in his own region. Against that is put the argument that it would be vastly expensive to drag all the panoply of authority such a distance. And so it would. Good point. In which case would it not be more reasonable to pay the expenses of the poor, miserable defendant in being dragged to London? But democracy does not work like that, and the M.M.B. would no doubt claim to be a highly democratic organisation. So the farmer has to pay his own way.

Although I was quite willing to go along with Watcyn to the Hearings Committee as some sort of barrack-room lawyer, by one of those quirks of fate I could not travel up with him because I was due to be in London for a meeting on the day before the hearing and a further meeting on the morning of the hearing. My meeting was in the Kingsley Hotel, somewhere in Bloomsbury. I cannot recall the name of the street, but I knew it was not far from the Russell Hotel.

Unfortunately, Watcyn had only ever been to London once and that was by train when he was quite young. So he knew nothing of the hazards of driving in the Metropolis when he knew not exactly where he wanted to go. Much of it is a question of getting into the right lane of traffic at the right time. It is generally accepted that it is not a good idea to be in the left lane, only to find when you reach the traffic lights that you want to turn right. A Londoner whom I know reckons that the best thing to do is to stay in the middle lane and, when in doubt, stop. His theory is that they will curse you and revile you and say all manner of evil things against you, but they won't run into you. Maybe that is so but, on the other hand, it doesn't exactly get you to your destination.

But Watcyn had a bright idea. There was ever such a nice chap who was a lecturer at some college in London and who often put his caravan on Watcyn's farm, and he had told Watcyn that any time he and his wife wanted to come to London they would be more than welcome to stay a couple of nights. No trouble to find their house. Not a mile off the M4 at Slough. I have to admit that it sounded like a good idea at the time.

I went on ahead to London leaving Watcyn with strict instructions to tell this lecturer to put them (his wife was going with him) on the nearest underground for Russell Square, from where he must take a taxi the short distance to the Kingsley Hotel where I would be waiting for him. Nothing could be more simple. I have never met the lecturer and even now, all these years later, hope I never do.

Now, it is not as idiosyncratic as it may sound, but quite frequently when I go to London I go via Birmingham. If you look at the map you will see it is about twenty-six miles further than the direct route by car. However, it is not all that joyful driving in London or trying to park or garage a car near where you want to go. Nor is the train journey cheap, and, at this end, it can be slow and tedious. I often found it convenient to travel from the bow-and-arrow country, here in the far west, by car to Birmingham, stay with a good friend, leave the car there, and go to London on the fast train service which runs between these two great cities. There was the additional advantage that the train came in at Euston, which is near Russell Square.

Thus it was that I found myself in London on the morning of Watcyn's hearing and waiting for him and his wife to turn up at the Kingsley Hotel by taxi at 12.30 p.m. That would give us time for something to eat and to discuss tactics. The hearing was at 2.30 p.m.

We can skip the paragraphs which could be taken up describing my nail-biting anxiety as the minutes dragged by. The taxi arrived eventually at 1.45 p.m.

Watcyn had indeed found the lecturer's house, not a mile off the M4, no trouble at all. But the lecturer had had a much better idea than mine altogether. He would accompany Watcyn and Ray in Watcyn's car, giving Watcyn instructions as to the correct lane in which to drive, and so on and so forth and find a place for him to park by the Russell Hotel. They reached the Russell without any trouble. Well, that's nothing to brag about with Watcyn having a college lecturer to direct him. But then, surprise, surprise, there was no place to park anywhere near the Russell. A college lecturer, not having to park near the Russell in the normal course of his duties, and only knowing where it was, couldn't be expected to know exactly where to park a car. So they drove a little way, seeking in vain for a place to park, and then they drove a little further, and a little further, until they finished up outside the lecturer's college and he knew where to park there all right. Well, he would, wouldn't he?

Then he put them on the underground and they did what they should have done in the first place. They travelled to Russell Square underground and presented me with some sort of problem. To start with, Watcyn didn't know exactly where the college was, but it was quite a big place and there were some railings round it. For that matter he didn't know the name of the college either, but in an effort to be helpful he said he knew the name of the underground from which they had embarked. It was Manor House. I looked at him in horror and said, 'Good God, man, that's out by the bloody Arsenal!'

'Is it?' he said.

You've read about Christopher Columbus no doubt. When he set off he didn't know where he was going, when he got there he didn't know where he was, and when he came back he didn't know where the hell he'd been.

It was a conundrum which would have to be tackled later. For the nonce, all that mattered was to have a quick snack and present ourselves at the Russell Hotel. Ray went off to do some shopping.

In addition to various manufacturers' literature, Watcyn was armed with a written statement from his vet saying, *'This is to certify that Mr W Richards of Bunkers Hill, Camrose, is a client of this practice and he has been advised to use, under our supervision, parentral terramycin as a treatment for mastitis in his dairy cows.*

He was advised to follow the manufacturers' instructions that accompanied the drug as far as dosage and the withholding of milk are concerned. These instructions do in fact state that residue in milk from parentral administration at recommended dosage levels are below the minimum level detectable by the standard (modified TTC) test.'

Who was there altogether I cannot recall and have no intention of trying to turn up any record of it. The two whiz-kids were sitting outside, having been brought there with all expenses paid. But I remember, because I knew them, that Sir Richard Trehane, then Chairman of the Board, was in the Chair, I think Steve (now Sir Stephen) Roberts was there, Stan Morrey was there, and so was one other Board member. I cannot remember what officials were in attendance, but they had a solicitor to present the case against the producer. If ever I am on trial for murder I hope he will be engaged by the prosecution. When it came to talking about farming matters he was still wet behind the ears.

At one stage he came out with the old nonsense that Watcyn was a most nefarious character indeed to have injected the cows himself and did he not know that only a vet was supposed to do that? Watcyn made as if to answer, but I kicked him under the table and suggested to the solicitor that he should address the same question to the gentlemen sitting at the top table. I smiled at the Board members, and they smiled back. They were all successful, practical farmers and, whatever complaints Watcyn may have about many injustices, including those suffered as the result of the incompetence of Board staff, he could have no complaints about the hearing that day. He was shown every courtesy and sympathy, and the entire proceedings were the essence of fairness. They knew as well as anybody what the extra cost to the already overburdened farmer would be if the vet had to be called out every time a tube of something or other had to be shoved up a cow's teat.

At the end Sir Richard Trehane said there were two of the Board's employees outside if we would like to call them in. I said, 'Do you want to call them in?'

He said, 'No, we don't want to call them in.'

So I said, 'Well we certainly don't. So leave the dull buggers where they are. They've caused enough trouble and confusion already.'

I only made one other point and that was that the late Lord Beaverbrook, who had not been without some small success in life, had always made it a point to sack the man at the top.

I heard afterwards that Sir Richard had been hurt by this remark because he thought I had been referring to him and he felt he had been very fair, which, of course, he had. What is more, I also knew what a tremendous contribution he had made to the farming industry. The next time I saw him I was able to explain that I had, in fact, been referring to the Board's regional manager to whom those lower down were answerable. Had he been doing his job properly the whole ridiculous business need never have happened.

The following day, June 17th, the Secretary of the M.M.B. wrote to Watcyn as follows:

I am writing to confirm that, at their meeting on 16

June 1976, the Hearings Committee decided, on behalf of the Board, that the agreement under which you sell milk to the Board should not be terminated. The committee further decided that, in view of the number of occasions on which your supply has failed to pass the test for antibiotics and other inhibitory substances, they may wish to reconsider the matter should a further test failure occur.

The committee wish me to thank you and your representative for your attendance at the meeting.

By this time Watcyn had worked out some of the answers for himself and there were no further failures.

Before that happened, however, we had to get home.

We finished at the Russell Hotel, as far as I recall, at about 4.0 p.m. and Ray was waiting for us with her shopping. As those who are familiar with London will know, it is a time of day which is known as the rush hour. It is a term which has always puzzled me. Many characters may be seen sitting in their stationary vehicles, presumably in the conviction that nobody will run into them. It is all rather confusing and frustrating. To travel by tube from Russell Square to Manor House was no more trouble for us than for the countless thousands of others who are daily subjected to such fiendish indignities. If farmers treated their cattle like it they would, quite rightly, be locked up.

Arrived at Manor House underground station we began to walk in the direction from which Watcyn seemed to recall having approached it earlier in the day. Eventually, on the other side of the road, I espied a greengrocer's with a splendid display of fruit outside. So I crossed over and bought two beautiful water melons. One was to take home, and one was for my mate in Birmingham. The character who sold them to me spoke a little English, but not enough to tell me how to find a college the name and location of which I did not know. The next four people I accosted also spoke little, if any, English. Maybe the sight of a chap with a thumping great water melon under either arm did not inspire them with much confidence.

Eventually, however, we came upon a gentleman, and I use the word advisedly and lovingly, who appeared to be the owner of a large shooting-brake from which he was delivering kegs of beer to the back-door of a pub. A splendid man he was. He not only spoke English, but explained that there was more than one college in North London. There seemed to be several of them. From Watcyn's description he hazarded a guess that the one we wanted was maybe a couple of miles in the opposite direction from that in which we were walking. So he gave us instructions and we set off. Ten minutes later this gentleman overtook us, pulled up, and told us to get in. Had he not done so we would never have found the place. If he should ever chance upon these words I would like him to know that his kindness will never be forgotten. His reward will be great in Heaven.

I was just beginning to entertain some doubts when Watcyn said, 'There's my car!' And in that moment I knew how the hard-pressed and confused Vasco da Gama must have felt when he gazed out upon whatever it was upon which he did gaze out. 'O Paradiso!' Notwithstanding which we still had a fair bit of navigating to do.

Maybe you would have some good ideas as to what directions to give to a country boy on how he should, first of all find, and then drive along, the North Circular road, at that time of day, from somewhere in North London to Slough. I entertained no such foolish notion as even to try. Possibly I could have navigated the journey at a more civilised time of day but, even if successful, I would have given little for my chances of returning from Slough to Euston to catch a train which would get me to Birmingham in time for the meal which I knew would be waiting for me. And you may recall that that was where my car was.

Having had a successful hearing in the afternoon, Watcyn now had even more faith in me than after the successful conclusion of the pollution episode. He took it as gospel when I told him that the best plan now would be for him to drive me to Birmingham, for I knew I could direct him from there. Even when we passed Holloway prison for the third time he was not one whit perturbed, but, at that point, I suggested we should perhaps move over into another lane and see what happened. It worked like a charm and, in no time at all, we were heading north.

We reached and crossed Birmingham safely and, arrived at the western fringes of that great city, I left Watcyn and Ray where I knew they could have a good meal, what time I went on to the repast that awaited me. I had given them detailed instructions for the route they should take and, in the event, they were home at Bunkers Hill just after midnight. I wouldn't say that was too bad, would you?

CHAPTER THREE
Cover-up and apathy

Back on the farm, and with things returning to normal, Watcyn brooded on what had happened and on other things which were still happening, and the resentment smouldered. He had been penalised financially by way of deductions from his monthly milk cheque, stopped at source without the option, and the jaunt to London had cost him money as well as time, which is worth money.

From time to time he has seemed to harbour some odd notion that there is such a thing as justice in this life. More than once I have reminded him that, once upon a time, there was a much better character than he will ever be, and all that society and the system did for Him was to nail Him to a tree. And who is Watcyn Richards to expect preferential treatment in the face of that?

His resentment eventually led him to take such action as would result in confrontation with the Min. of Ag. and Fish., the uselessness of some of whose officials has been referred to in the first chapter. Such episodes will be dealt with in chronological order, but it is as well to remember their presence at all stages.

Even the red rose Socialists these days seem to be keen to shed their image as being a party of high taxation and to concentrate instead on promoting the idea of the great advantages and benefits of increased government spending.

But how many people stop to consider how little of the tax collected seeps through for the public's benefit? I read somewhere the other day that the figure is no more than five pence in the pound, whilst the vast bulk goes to maintain the administrative machine with its ever-growing hordes of public employees. It can hardly be long before the number of public employees will outnumber the rest of us, and even Mrs Thatcher was impotent in the face of such a tide surging against her. Lip service to the good cause continues to be paid from time to time but the public payroll still increases inexorably.

What is even more depressing is to come up against examples, time and time again, of the inadequacy and incompetence of so many who are grossly overpaid for what little contribution they make to the common weal. The pathetic feather-bedded inadequacy of some of the misfits in local government beggars description. And the same can too often be said in the case of so many Boards, Unions, Commissions, Trusts, Societies, Corporations and all the rest of them.

Nor is it only a matter of the incompetence and inefficiency. There is also the attitude of stuff-you-mate, and pull up the ladder, Jack, I'm on board. Indeed, this aspect, which is symptomatic of the age in which we live, is perhaps the worst feature of all.

Still, as mentioned earlier, the M.M.B. did eventually introduce the system of labels for the milk sample bottles, and then Pfizers announced revised withholding times following the use of Terramycin. They subsequently asserted categorically that this in no way suggested that they had found anything faulty in their earlier recommendations. And the band played, 'Believe it if you like.'

For my own part I had seen enough to be deeply

concerned, not only for Watcyn, but for dairy farmers generally, although I had by that time been out of milk production for about six years.

I told the story of my own losing battle with the minions of the Min. of Ag. and Fish. in my book 'What Price Abortion?', and I knew something of the brick wall against which Watcyn now proposed to bang his head. In the same book I also referred to the apathy and smugness of farmers who were not aware of what was going on around them.

To me, it seemed that the obvious course for Watcyn to take was to enlist the support of the National Farmers' Union of which he had been a fully paid-up member since the time he had started farming. So I told him to take it up with his local branch and have a resolution sent to the milk committee of the County Executive.

In the meantime, and it is perhaps as well to deal with events in chronological order as far as possible, Watcyn had another milk failure. That was in January 1977. Fortunately more than six months had elapsed since the previous fandango, so he was now starting from scratch as it were.

At this stage I told him to put everything in writing and, where it was unavoidably essential to telephone, then make a note of the conversation at the time. The telephone is a wonderful servant, and maybe quicker and cheaper than post for anyone who can't afford a Fax. But it is all too tempting for the Watcyns of life, on receipt of another screed of cretinous humbug, to pick up the telephone and blow off steam. Worse still, the characters at the other end can so easily come the old pals' act, 'quite agree with you personally but it comes from higher up and leave it to me and I'll sort it out for you.' And, once again, the only word that comes to mind is 'Cobblers'. Let us have it in writing.

Wherefore and therefore, Watcyn started the new episode by writing two letters. The first was to the Head of the Milk Quality Department who had notified him of the recent failure and who had enclosed two new screeds from the M.M.B. explaining how many beans make five. One was MQ244 which now went so far as to warn that antibiotic residues could remain in the milk for 'up to about fourteen days' and advising that milk from treated cows should be withheld for seventy-two hours, or six milkings. Far from telling Watcyn anything he did not know, it was what he himself had already been trying to tell anybody who was willing to listen to him, but at least it was progress of a sort.

The other screed had the reference number MQ285 and said, *inter alia*, 'The producer will normally receive an advisory visit from a number *(sic)* of the Board's Regional staff.'

Maybe it was a misprint, but Watcyn was not to know that, so he said in his letter, 'As a matter of interest I would be glad if you could tell me whether this means that they are likely to descend in ever increasing hordes.'

He also enclosed a copy of his second letter, which was much longer, and addressed to the Regional Manager of the Board. In it he explained that he was putting everything in writing so that, in the event of any fresh nonsense arising, there would be no doubt or argument as to who had said or done what on the various occasions. He emphasised that he had again faithfully followed out the instructions on the manufacturers' label and added that the relevant details would be produced and an issue made should the need eventually arise.

He also mentioned that, two days after the unsatisfactory test, he had received a visit from one of the two dynamic whiz-kids who had journeyed to London the previous

June, all expenses paid, without even having been asked into the room, and who had now come at the behest of Kraft (the eventual buyers of Watcyn's milk) to take a sample on their behalf. The letter continued:

If he were capable of reading he could have seen from the M.M.B. instruction chart, to which I drew his attention, that there was a strict procedure for the taking of milk samples. These instructions he deliberately ignored and took the sample when the milk was still at 67 fahrenheit.

This, then, is the calibre of the personnel who would advise and help me in the unlikely event of my being sufficiently misguided as to submit myself to their idiot ministrations.

Watcyn, the official admitted, was the only producer being singled out for his attention although he said there had been a dramatic rise in the number of failures in the area. He concluded his letter by saying, 'I trust you will keep yourself abreast of these developments because I want to assure you that this time you are called upon to answer for the purblind stupidity of those under your control.'

For some reason, Watcyn forgot to mention that, instead of storing the sample in an ice-box, as per instructions, the whiz-kid stuffed it into a shoe in the boot of his car.

The reply to this letter was another helping of incredible waffle so that Watcyn had to write back and, in his opening paragraph, said, '... I have come to the conclusion that you are no more fit to be doing the job that you are paid to do than those who are answerable to you.'

There were a few other such pleasantries but this letter did at least elicit the positive reply, '....it is not our normal policy or practice to sample on behalf of Krafts or any other Dairies, but should circumstances arise when we think this may be helpful then the Board's field staff are fully authorised to do so.'

We were not destined to learn what circumstances had arisen to suggest that this particular exercise would be helpful because the case died on its feet. There were no further milk failures for the time being and, as far as the Board were concerned, for a while a great peace descended on Bunkers Hill. Watcyn did, however, at my instigation, send a copy of this correspondence to the late Jim Morton, the Managing Director of the M.M.B. He was an old, personal friend of mine and a first-class man in every way. Watcyn concluded his letter to him, 'I want everything on record this time as to who said or did what and when, for if I am ever again subjected to the same as I had to suffer last year as the result of the incompetence of Board staff in this area, I am determined that heads will roll.'

Whilst all this was taking place Watcyn was following up my suggestion with regard to the N.F.U., and his local branch duly sent a resolution to the County Executive for consideration by the Milk Committee. I had been a member of this committee for a long time and, exactly twenty years previously, had been its chairman. I had some foolish notion that my views may have been worth something, particularly as I was already starting to be proved right in many of the things I had been saying in previous years about brucellosis and the useless S19 vaccine being peddled by the Min. of Ag. and Fish. Yet, remembering the pathetic indifference and myopic self-interest of so many of the members over that particular issue, I should have known better.

How well I remember one particular meeting when I was trying to open people's eyes and was prepared to quote my own bitter experiences, and one wretched little man got up and whined, 'Mister Chairman,' (and how I wish I could write it as he said it) 'we must be careful about the publicity over this sort of thing because it will affect the price of our cows in Carmarthen mart.' Never mind, he was a Tory sycophant for long enough to ensure that he climbed to the dizzy heights of appearing a couple of times in the biannual farce of the Honours List.

Amidst all the high-faluting talk the only real consideration was how to divide the money between the successful and the fortunate, with nobody troubling a couple of monkeys' uncles for the poor so-and-so who was in trouble and really needed the financial support. 'For unto him that hath shall it be given, and he shall have abundance: but from him that hath not shall be taken away even that which he hath.'

So many of them could not see that they, too, were just as vulnerable to the hazards inherent in the system, and were smug enough to fool themselves that it was their good management which had kept them out of trouble.

Years ago, I recall a Brains Trust at our local N.F.U. branch and the question was read out, 'If one sheep is another sheep's worst enemy, who is another farmer's worst enemy?'

The audience chuckled and the members of the Trust didn't even try to answer. And there is hardly a farmer in the land who doesn't know the answer.

When the resolution from Watcyn's local branch, where members had shown every concern and sympathy, came before the Milk Committee, the chairman of the local branch, who should have made the case, got up and said, 'Ah well, then. It's come from my branch, so I suppose I'd better propose it,' and sat down.

As the discussion continued I became more and

more disgusted. It was the last meeting of the County N.F.U. I ever attended. I still show my face occasionally at meetings of my own local branch, but that is only by way of meeting up again with a few old friends for old times' sake, and I think I am wise.

When I was younger I used to listen to some of the old jokers carrying on in committee and I used to ask myself what it had to do with them when they were no longer producing milk or even farming. So maybe I was now in the same position. I was sad, though, that they had no more interest in the case of a younger member who was in dire need of their support. Although I am a life-long member of the N.F.U. I could understand Watcyn's attitude when he eventually cancelled his membership.

Before that happened, however, his Group Secretary did his best for him and was quite splendid. He could not say much, but I know he thought his own thoughts about the way Watcyn had been let down by his fellow members.

The Group Secretary prepared a detailed report of the case and sent it to Pfizer suggesting that they should at least pay Watcyn's out-of-pocket expenses in respect of his totally unnecessary trip to London. He had, after all, used their product which, according to the trial data and leaflets issued with the drug, would not leave residue in the milk detectable by the standard (modified T.T.C.) test.

The reply was predictable, and even Watcyn saw no point in pursuing that one.

CHAPTER FOUR
A difference of opinion

Huckleberry Finn said, 'If I'd knowed what a trouble it was to make a book I wouldn't have tackled it and ain't a-going to no more.'

Maybe, like Huck, I am beginning to think that I should have felt a good deal less sanguine when I embarked on this effort and have already reached the stage when I am wondering where on earth to turn next. I have just been reading yet again the letters and reports and articles on which the next chapters have to be based and, the more I read, the more the mind boggles, for there is enough scientific detail for a learned treatise.

Apart from the fact that I would not be qualified to write such a book, the interest would be strictly limited. On the other hand, the incidents involved lead up to other wider issues and it will be necessary, therefore, in discussing them, to know something of what went before. A certain amount of background information is essential.

[I have given a great deal of thought to the desirability of condensing this chapter, which deals mainly with the misuse of antibiotics, but there are those concerned with the issue who have read the chapter in manuscript and who have insisted that it is of great importance. Those who have no interest in the subject need not read it.]

In addition to all this there were several other issues

developing at the same time, and to some of them it will not be possible to make more than passing reference, if that.

Some wise character once said there are three sides to every story. There's your side, there's my side, and then there's the truth. Granted it could be said that Watcyn is pig-headed and stubborn or, as we say in Pembrokeshire, awkward. That is not my experience of him, but I can understand that others may have found him so. At any rate, he was, by now, spoiling for a fight, and it was another case of an irresistible force being met by an immovable object.

Nine people out of ten will say that Watcyn should have allowed the matter to rest, and nine people out of ten could well be right. But then, that would not have been Watcyn. It would have been foreign to his nature. It was enough for him to see a touch of incompetence here, a few falsehoods there, and a measure of self-interest, plus all the usual ingredients, and his Celtic blood was up. These days they say the adrenalin was flowing. It will be a sad day for society when there are no longer those who are prepared to challenge 'the system', although any sensible person will tell you that you can never beat it. Watcyn was still at that time labouring under the delusion that there was such a thing as justice.

Not being able to find any support, or what he thought to be sense, from any of those from whom he would have expected it, he then decided to carry out some tests of his own. All he proved to his own satisfaction was that it was no use following the instructions on the labels too closely, and there were some more notifications of antibiotic failures as proof of that. So then he tried feeding cider vinegar. The results, at first at any rate, were encouraging, but whether it would have been any more beneficial than feeding the raw apples to the cows is something which could more usefully be discussed on some other platform.

The hunt now began to open up on all fronts with Watcyn firing in several directions at the same time. It is not perhaps the surest way of hitting the target, but highly diverting to the interested onlooker, and there is always the chance of the odd shot finding the mark and causing a few casualties as the quarry is flushed from cover. One day I must draw up a list of how many of them decided to take early retirement after Watcyn started his campaign, and then perhaps a supplementary list of how many would dearly have loved to.

It was in the summer of 1981 that the scarf again came into the reckoning.

The trouble started with an article in the *Farmers' Weekly,* as a result of which Watcyn was sufficiently misguided as to telephone the local Trading Standards Department. The letter which arrived in response to this call said, *inter alia:*

This Department has no power to interfer (sic) in the levels of antibiotics acceptable by the Milk Marketing Board or to enquire into the sampling proceedures (sic) regarding milk sold to the MMB, this is surely a subject of your contract with the Board and should be discussed with them.

If having done this you are still agrieved (sic) by the replies received, you could contact the Minister responsible for agriculture either directly or I would believe more preferably through your union, with the assistance of your member for parliament.

A further possibility would be to take samples of your own at the same time as the samples are taken by the MMB and to submit these to the Public Analyst for independant (sic) analysis, this could help you decide whether the drugs are at fault or the sampling and analysis proceedure (sic) of the MMB.

To put it another way, they simply didn't want to know.

We have already discussed the Union aspect and, as far as the then M.P., the then Rt. Hon. Nicholas Edwards, was concerned, he was also by now the Secretary of State for Wales, and therefore the satrap with overall responsibility for much of what was happening in the department of agriculture. The spelling in the letter is just another miserable reminder to an older generation of how the taxpayers' money has been squandered on what passes for education in turning out a nation of semi-literates.

Pressing on regardless Watcyn carried out a few more tests to prove his point to his own satisfaction, which he did by having some more notifications of antibiotic failure. For my own part I cannot for the life of me imagine what satisfaction he could possibly have derived from all this, unless it could have been the re-emergence of the scarf. Not that the scarf appeared on the scene in person. No doubt being mindful of the dire threat concerning his scarf and the adjacent tree, and probably taking it literally and seriously, he sent along, in his stead, his side-kick, earlier identified in this narrative as the other whiz-kid.

Samples were taken on behalf of Kraft Foods, after the recommended period of withholding of milk had elapsed. The morning sample failed. The following morning it passed. Any reasonable person would have assumed that the coast was now clear. But nothing of the sort. The following evening came up with another failure.

I am not conversant with the position these days, because modern technology is causing changes and advances at a bewildering rate, but those who should know tell me that the position about withholding times, not only for milk but for animals going for slaughter, is more chaotic than ever, with yesterday's instructions being contradicted tomorrow, and the Veterinary Products Committee being a law unto themselves.

And, in the face of all this nonsense, it is the farmer's responsibility and he has to carry the can. Then they bombard him with literature imploring him to call in advisers on whom there is no more dependence than on a baby's bottom.

This was highlighted and emphasised by an article in the *Milk Producer*, the M.M.B.'s official publication, in September 1981, where it said:

... withholding times for milking cows that are being treated for mastitis are given both on the tube and the box. But under the MMB's sale conditions, it is the milk producer's responsibility to ensure that the milk which he offers for sale is free from antibiotics or any other inhibitory substance.

It did nothing to improve Watcyn's humour and he wrote a letter which duly appeared in the *Milk Producer*.

The article on antibiotics in the September Milk Producer implies that the Board are anxious to help farmers who are experiencing problems. The first thing they should do is ensure that the writer of the article is locked up out of harm's way so that he can no longer mislead poor fools who know no better.

The only glimmer of sanity in the article is where he says: 'The majority of test failures are the result of accidents.' Even this is only partly true. The real truth is that the failures are caused because the producers are being hopelessly misled.

In a brief letter such as this it is is possible only to generalise.

The letter then went on to spell out the two different standards regarding the withholding period as already dealt with in these pages, and concluded by saying:

In fact, the inhibitory substances can be often still detected many days after the recommended period for withholding has elapsed.

From the MMB's screeds of hogwash it is evident that they are fully aware of this, but they do nothing about it because they serve not the dairy farmers, as they claim, but the hordes who live upon the farmers' backs.

The letter was published, but the Editor also included beneath it a lengthy reply from the Director of the Board's Technical Division. Whereas Watcyn's letter ran to forty-one lines, the Director's reply ran to ninety-six. I do not find it to be of sufficient interest to quote it here, but it said nothing at all on the vexed question of intra-muscular injections, although it dwelt at great length on the so-called dilution factor and intra-mammary treatment.

Quite reasonably, Watcyn felt that this reply, alongside his own letter, called for further comment. He had an acknowledgement from the Editor saying that this letter would be published, but it was not. In the fulness of time he was fobbed off with the comment that as one letter had been published he should be satisfied as other producers also wished to express opinions. This is the price of democracy and, as I mentioned earlier, the Board is a democratic organisation. Or supposed to be. I think.

Watcyn's second letter had said:

Following my letter in the October Milk Producer, and the reply to this by Mr Frank Harding, Director of the M.M.B.'s Technical Division, the thought occurs that use of the term 'hogwash' may have been more appropriate than on the surface it appeared to be.

Mr Harding made no reference to intra-muscular injections and these, too, can be the cause of much trouble and heartache.

Readers must form their own assessment of his reply. This apart, could Mr Harding now answer a few more questions?

If, according to the Board's hogwash, contaminated milk should be fed to stock, why can it be assumed that this milk, which is harmful to humans (which is not in question) is not equally harmful to stock?

When contaminated milk is fed to pigs, some of which are already being given feed which contains antibiotics, what is the likely effect on the pigmeat?

What is the effect on the unborn calf when the pregnant cow is treated with antibiotics?

To what extent is the increasing inability to treat calf scour directly related to the resistance to antibiotics which has been built up, either by treating the pregnant cow, or feeding contaminated milk to the calf, or both?

If Mr Harding is not in fact the fount of all knowledge on the subject, is there anyone else belonging to the Board who would make so bold as to offer an opinion?

Both as a farmer and as a journalist I would have said that the answers would have been of tremendous interest to readers. Such, however, is the price of democracy and the more is the pity. Either the Editor, at that time a Jenny Bradley, or those above, decreed that the readers, milk producers all, should be kept in the dark. Later on, however, the advice to feed contaminated milk to livestock was quietly dropped from the Board's advisory literature, but I don't suppose that it had any more to do with Watcyn's protestations than Pfizer's change of advice in respect of withholding times. Nothing at all. Of course not.

About the same time, there was reference in the *Farmers' Guardian* to an appeal from the M.M.B.'s Technical Director for 'details of preparations used for intramammary treatments from which inhibitory substances can often be detected many days after the recommended period of withholding has elapsed.'

Watcyn, therefore, wrote to the Editor and said:

Firstly can we nail Mr Harding on his insistence on referring only to intra-mammary treatment? Intra-muscular injections for foul in the foot are just as lethal, and of just as much concern to the milk producer because these, too, come up with the dreaded inhibitory substances.

If Mr Harding wants evidence, let him read articles which have appeared on the subject in national agricultural journals.

When he has done that, let him look into my own case history, and, if he can arrange with the Board's local staff to accommodate him in between their holidays and weekends, let them carry out tests in this herd at somebody else's expense instead of mine, for a change.

There was no reply and there were no takers.

Not for anybody to feel left out, or to think that Watcyn was being too selective in his approaches, he also paid a visit to the office of the Divisional Veterinary Officer, Mr David Russell, who subsequently wrote to say:

Since your recent visit to this office and our discussion on antibiotics in milk I have enquired about the latest position with regard to the withdrawal period for intra-mammary preparations.

The Veterinary Products Committee and the Milk Marketing Board are very concerned about this matter and have had discussions to decide on an agreed and an acceptable level of antibiotics in milk. It is difficult to say when these discussions will be completed and when action will be taken as a result but it is hoped that this should be early next year.

Going through the files I have come to the conclusion that the ensuing winter would seem to have been fairly quiet. Perhaps Watcyn had to spend a fair bit of time at choir practice because a man has to get his priorities right.

However, by the spring of 1982, it had become apparent from the prevaricating reply of the Editor of the *Milk Producer* that Watcyn's second letter, to which I have already referred, was not going to be published, and so he addressed the same questions by way of a letter to the editors of various farming publications. As a result, in the *Farmers' Guardian* of May 28th, there appeared a letter from a Professor John R. Walton, of Liverpool University, who said:

Mr Watcyn Richards says that milk containing antibiotic residues is without question harmful to humans.

May I ask from where does he get this information because we know that probably only one class of antibiotic when present in milk may cause significant problems to a specific group of people but this certainly does not apply to the bulk of antibiotics we use in agriculture and there are ample data to support this.

Secondly, your correspondent asks what is the likely effect on pigmeat from feeding antibiotic contaminated milk.

There will be no deleterious effect on pigmeat from such a practice even if other antibiotics are being fed at the same time, and if common sense is used in adhering to specific withdrawal periods then neither will there be any hazard to man.

As far as the unborn calf is concerned, as it does not

pick up any bacteria – which will ultimately colonise its intestinal tract – until it has been born, then any antibiotic circulating in its dam's blood will be irrelevant as far as failure to treat calf scour is concerned and in any case, many types of scour are not caused by bacteria therefore antibiotics should not be used in treatment.

One point that must not be forgotten is that milk is also used to produce cheese and yoghurt and the special bacteria that are employed to make these two products are very sensitive to the presence of antibiotics in milk.

Let us not forget that many millions of doses of antibiotics are taken by mouth by humans every year with no harmful effect at all and the amount being consumed at any one time is many times greater than would ever be present in meat or milk.

There are, however, certain rules to follow if these same antibiotics are to retain an effective life-span, and not fall into disrepute because they are mishandled by persons with no knowledge of their properties or correct application. Unfortunately, too many people feel that antibiotics are a panacea for all their problems.'

For my own part I would say that many millions of doses of antibiotics are also taken by mouth by humans every year with incalculable harmful side effects which may or may not be evident at the time. What is more, I speak with some feeling on the subject as a result of having been stuffed full of streptomycin when I had brucellosis, and which permanently damaged my hearing so that I now have to wear a hearing aid.

Even though streptomycin may not be classified as an antibiotic, it would certainly come under the heading of panacea so popular at the time, and without any thought of the side effects any more than with the antibiotics. And they gave me my fair share of those as well.

Watcyn, however, followed this letter by writing:

Dr Walton could not have been more fatuous and misleading had he been deputed to reply on behalf of the M.M.B.'s Technical Division, but there is a world of difference between theory and practice.

Suffice it for me to quote from a joint statement issued by the British Veterinary Association and the Royal College of Veterinary Surgeons when commenting on the Swann Report in 1977:–

'Over a considerable period of time the practice has become established of feeding antibiotics at sub-therapeutic levels to groups of animals which, although not at the time diseased, are entering a period of challenge in which experience has shown that an unacceptable incidence of disease of sub-optimal level of growth can be anticipated.

The committee is doubtful of the value of this practice: it is, however, in no doubt regarding its general undesirability and it finds that 'the present practice of giving antibiotics below therapeutic levels is indefensible from a bacteriological standpoint.'

If, by any chance, Dr Walton is seeking to imply that antibiotic residues in milk are not harmful to humans then what does he suggest all the fuss is about?

Understandably perhaps, Dr Walton would appear not to have cared overmuch for Watcyn's letter, for his next letter, which appeared in the *Farmers' Guardian* of July 23rd, said:

From W. Richards letter it would seem that my comments struck a tender spot, and contrary to his implications I hold no brief whatsoever for the MMB.

As a practising veterinary surgeon I consider

that antibiotics are perhaps my best weapon against many diseases affecting our domestic and food producing animals and I demand that these drugs are used by those people who know how to use them properly.

The effective use of these sophisticated drugs is not by DIY but by those with full knowledge of their properties if antibiotic tissue residues are to be avoided.

It is unfortunate that Mr Richards chose to quote from a rather out-dated joint statement. Knowledge has advanced rapidly since 1977; for instance, at an open meeting in London in September 1981 it was agreed by the delegates which included the deputy chairman of the Swann Committee of 1969, that some evidence provided for the Swann deliberations had been misleading and this resulted in confusing statements about the low level use of antibiotics in farm animals.

Finally – Why are antibiotic residues in meat or milk not wanted? – because (a) legislation decrees it, (b) manufacturing processes for cheeses etc. are compromised and (c) the presence of one specific antibiotic may present a health hazard for a certain group of people.

As it happened, Watcyn received a letter round about this time from the Veterinary Products Committee at Weybridge. He therefore sent a further letter to the *Farmers' Guardian*:

If Dr Walton reckons that I am out of date in my facts, may I bring him up to date by quoting from a letter written to me as recently as July 27th this year on behalf of the Veterinary Products Committee at Weybridge:–

'Problems arising from the existence of differing standards for acceptable residues of antibiotics in milk are being considered as part of the review of all intramammary products licensed under the Medicines Act and

the Veterinary Products Committee has a major task in trying to reach a solution which will meet the essential needs of the dairy farmer and be acceptable to Milk Marketing Boards, the Local Authorities and the various manufacturers who hold the product licences.

I can assure you that the particular aspect to which you have drawn attention is not being neglected and you will be interested to know that in addition to the standards you quote, 0.05/0.02 international units of penicillin per millimetre of milk, the Committee is having to consider suitable standards for all the other dozen or so antibiotics which are used in intra-mammary preparations.'

I can find nothing else in the file, so I am not sure whether the *Farmers' Guardian* declined to publish or whether the Professor withdrew from the correspondence.

Some time later, in the January 1988 edition of the *Milk Producer* a letter appeared from a lady who referred to the misfortune experienced when water accidentally entered their milk, and they duly reported the fact to the Board, only to discover that insurance cover was against antibiotics only and not water.

In reply the Editor explained the position and went on to say, *'Antibiotics are a different case because of the risk to public health ....'*

I wonder whether Professor Walton was ever in the habit of reading that publication. If so, it would have been kind and helpful of him to tell them all about it.

CHAPTER FIVE
Some have to stand alone

Although so many other issues were developing at the same time, it is as well, because of so much which was to come later, to confine ourselves for the moment to the question of milk tests and authorities various.

In March 1982, Watcyn received a communication from the M.M.B.'s Central Testing Manager for the region. It notified him that the Board had decided to centralise the testing of ex-farm milk for quality payment purposes through six Central Testing Laboratories, to test for compositional quality, hygienic quality and screen for antibiotics, on a weekly basis.

Labels for Watcyn's use in co-operating in the trial were enclosed. In the light of all that has gone before it will come as no surprise that he had no intention of co-operating, so telephoned the gentleman in whose name the letter had been sent and told him so. He then wrote, confirming his telephone call:

I am in receipt of your letter of unknown date, and it is of no use heading it, 'Date as postmark', because the postmark is indecipherable. To put it another way, it is no more accurate than your tests.

I wish to confirm what I told you on the telephone on March 8th.

My views are already well-known to the M.M.B. and

it is no use for the Board or anyone else to talk of more stringent standards applied to producers in European countries.

We are having to use products which have been taken off the market in Ireland and on the Continent. The drugs we are having to use are cleared by the Ministry of Agriculture to a standard 0.05 but the Board test for inhibitory substances to a standard of 0.02. If the manufacturers' instructions are carried out, there is bound to be trouble for the producer, because these instructions cannot hold good when the Board are testing at a different level from that for which the instructions are intended.

Do you understand this? If not, you must be obtuse. If you do, you will realise why I have no intention of co-operating in this new scheme blatantly implying approval of such a travesty of justice.

Inevitably there were letters pointing out Watcyn's contractual liability and issuing certain dire threats. These fell slightly short of having him shot at dawn, but that, as the final solution, might have saved a great deal of trouble.

Undeterred, and for good measure, Watcyn also made it known to the Min. of Ag. and Fish. that he now intended not to co-operate with them either in the carrying out of routine blood tests for brucellosis and tuberculosis. This elicited a reply from the Veterinary Officer, Mr Q.C.Wadsworth, who said:

From our telephone conversation today, 2 July 1982, I note your demand for co-operation from Welsh Office in your dispute regarding failures to the M.M.B. tests for inhibitory substances in milk, and that in persute (sic) of your demand you are prepared to refuse Welsh Office staff access to your farm to complete routine testing procedures.

As discussed I must again confirm that it is inappro–

priate for this office to write any claim on your behalf and therefore I must again request that you, or a representative writing on your behalf, write formally setting out the grounds of your dispute so that it may be given due consideration.

If it will be of any assistance any such letter may be sent to this office for onward transmission to the appropriate body which in this case I understand may be the Veterinary Products Committee.

I have already referred to the appallingly low standard of literacy of so many being turned out today by what passes for an educational system, so that it is not surprising that one of its products should write 'persute' for 'pursuit'. Though the word cannot be confused with 'dispute', or even 'hirsute', I can find it in my heart to forgive the one who has dictated it for not having noticed the error when it comes to signing it.

But I find it hard to forgive anybody who can talk about 'onward transmission'. Presumably he meant he would forward it or simply send it to.

Watcyn replied as follows:

Thank you for your letter of July 2nd.

If you will refer to a letter written to me by Mr David Russell on Oct 21st '81, under the reference number 909, you will see that I was led to expect action early this year.

In case the hen has died on the nest before anything could hatch out, may I repeat in words of one syllable that I am sick to death of pointing out that the drugs we are having to use are cleared by the Ministry of Agriculture to a standard of 0.05, but the M.M.B. test for inhibitory substances to a standard of 0.02. If the manufacturers' instructions are carried out, there is bound to be trouble for the producer, because these instructions cannot hold good when the Board are testing at a different level from that for which the instructions were intended.

If you can now put this into an onward transmission situation, or any other jargon you can think of, it will be only a matter of years later than when you should have pulled your finger out. I regret I am not familiar with the jargon for that expression.

It would appear that this letter was duly put into an onward transmission situation, because it was in reply to it that Watcyn heard from the Veterinary Products Committee, and it was from that letter he quoted in his final reply to Professor Walton.

I have already referred to my own disenchantment with the N.F.U. in the failure of members to support both myself, originally, over brucellosis, and, of course, Watcyn. He was involved at the same time in a protracted dispute over his tractor and safety cab, which will eventually need a chapter on its own. It has to be said that, in the case of this dispute, both the County Secretary, Patrick Edgington, and Watcyn's Group Secretary, Keith Bevan, did everything they could to help, and he put his appreciation on record in a letter to the County Office. But what Watcyn needed was the backing of the Union at the top level, and this could only have come from a resolution at County level.

Earlier in the year, in January 1982, the then President of the N.F.U., Sir Richard Butler, had been due to visit the county. Watcyn wrote to him and asked if he could meet him, but he received a reply from Sir Richard saying that he was quite satisfied that the N.F.U. had done everything they could possibly have done to help him, in the business of the tractor and safety cab, and no meeting, therefore, was arranged.

In his letter Watcyn had said that he had been a member of the Union since he had started farming thirteen years previously, had done all his insurance through the N.F.U. Mutual, and would be reluctant to turn to the F.U.W., but he was not prepared to let spite or personal vendettas stand in his way. So what could they have been? I can think of a few.

Not long after Watcyn had started farming, the County branch of the N.F.U., in their fight for a better deal, suggested boycotting the local marts for a week or two. I can't remember the details now, thirty years after the event, but it was something like that. Watcyn knew nothing about it and, in those early days of his farming experience, had not even been initiated into the mysteries of this great organisation. So, having had a cow calve, and needing the money, he took the calf to the mart. A prominent N.F.U. member was there on watch and told Watcyn he couldn't sell his calf. Sufficient will have been recorded about Watcyn's temperament so far in these pages for it to be evident that here was one enemy hereafter when anything to do with Watcyn came up in committee.

Another character, before Watcyn came to Bunkers Hill, had made a habit of allowing his sheep to wander withersoever they listeth, and when that happened to be Bunkers Hill he would drive them back over the hedge which they had already flattened. When Watcyn explained that this was no longer permissible and obliged the gentleman to drive the sheep the long way back round the road for the second day in succession he decided to sell the sheep. Not a man who could be guaranteed to support a chap's cause in committee.

Worst of all, however, must have been the occasion when Watcyn took issue with the Hunt. He sent word to them in the morning that he did not mind the hounds crossing his fields or the horses trotting along the headlands, but he did not want them to gallop across his fields. But they heeded not his polite request. And he had seeded out some pasture and put up electric fencing. And the horses galloped over the new seeded pasture and broke down the fencing of this mere tenant. So Watcyn went out with the gun, fired two shots in the air to bring the horses to attention, and then rounded up the whole circus into the yard and kept them there for two hours.

The whipper-in, if I remember rightly, was smart enough to escape, but he and the horse went arse-over-head as they tried to jump the river. Well, the bank gave way didn't it, and there was the poor chap's hat floating down the river and maybe he would have been better off if he'd stayed with the rest of them, and the dogs had scattered to the four corners of the parish anyway.

I know two hours is a lengthy penance but, in all fairness to Watcyn, he would have settled for an hour. Just as he was thinking of relenting, however, one of them started to give his cheek and throw his weight about, so Watcyn offered to drop the horse under him and deal with him personally afterwards, which put rather a different complexion on things. And at that point Watcyn decided that one more hour would be just about right to allow him to cool off. Eventually, as the shades of night were falling, he let them out one at a time so that he could count them. Thirty-six of them there were, and some of them would have had no intention of voting to support Watcyn in any cause whatsoever, no matter how just that cause might have been. But, fair play, they paid up without argument for the damage done by their horses.

This episode, of course, will have written Watcyn off as a complete outsider in some folks' reckoning, but it will have endeared him to the hearts of others.

One prominent hunting member of the N.F.U., however, and not a bad friend of mine at that, assured me, when in his cups, that he would do everything in his power to ensure that the Union didn't lift a finger to support Watcyn So-and-So Richards. He did, too.

There was also the character who had been responsible for causing Watcyn such horrendous losses by pollution of the river, as recounted in the first chapter.

So much for the reference to vendettas. Then there was the reference to the F.U.W.

Those outside farming may wonder who they are. The initials stand for Farmers Union of Wales. It was founded, if that is the word, in 1954 by the then disgruntled Secretary of the Carmarthenshire Branch of the N.F.U., J.B.Evans, along with a handful of nonentities who were quite happy to become very moderate sized fish in a very small pond.

I regarded that organisation then, as I still regard it now, nearly forty years later, as a complete and unmitigated disaster. There were those, who knew no better, who said that it was a good thing because it put new life into the N.F.U. and helped to keep them on their toes. The opposite was the case, for it took the backbone out of the N.F.U. in Wales. Genuine people, who could see much that was wrong with the N.F.U., and there was plenty, were afraid to to say so because they were immediately told by reactionary old fogies that it was rebels' talk and playing into their hands.

At the time I was writing a country column under the pen-name Barn Owl and I referred to this new organisation as J.B.Evans' outfit. J.B. didn't like that at all and, at a subsequent public meeting, he kept expostulating, 'We are an union! We are an union!' Say it out loud a few times and you will realise why they soon became known as The Onion.

The harm they have done to Welsh farming is incalculable. Who was it said, 'Divide and rule'? Times without number the evidence has been there of people in every walk of life, with no love for farmers, who have exploited this division. They can be numbered amongst politicians, councillors, civil servants, journalists, broadcasters and anybody and everybody who has sought to divide the industry.

I said politicians. For years The Onion struggled on, claiming from time to time that they had been recognised when all they had been was identified. Then their moment came. Early in their existence they had appointed a young barrister by the name of John Morris as their deputy secretary and legal adviser. Then he became a politician as Labour M.P. for Aberavon. Then he became Secretary of State for Wales, and he recognised The Onion officially on behalf of the Labour Government. It was a terrible disservice to Welsh farmers.

Tragedy though it has all been for Welsh farmers, and although so many good and sensible people can see it, there is nothing anybody can do about it. All attempts have proved to be abortive. The unhappy truth is that The Onion are kept in being by the insurance company who backed them when they came into existence. Added to that is the fact that the smaller-minded regard chairmanship of various committees, no matter how insignificant, as Holy Cows. And Holy Cows, be they ever so useless and unproductive, are not slaughtered.

I have written of my aversion to The Onion for no other reason than to demonstrate what must have been the extent of my concern for Watcyn and the feeling of complete impotence when, in desperation, I agreed with him that he might just as well give them a try. Like the old lady said about the chip of wood in broth, 'There's not a lot of good in it, and not a lot of harm, but it's something to stir.'

There was one aspect in particular which prompted this approach. The President of The Onion, Mr T. Myrddin Evans, had for some time been expressing publicly his grave concern over the high mortality rate which had assumed alarming proportions in young calves.

It seems odd to me that in so much correspondence, and in so many utterances, so many people have hinted or implied that Watcyn was one of those with an irresponsible attitude to the use of antibiotics, whereas the opposite is the case. He had been warning of this sort of happening for a long time. Not surprisingly, he believed that antibiotics were now at the root of the present troubles, because calves with the dreaded white scour seemed to have an in-built resistance to the antibiotics previously used in the treatment for white scour, and in some cases could have been dying of antibiotic poisoning anyway.

So it was that, to my own amazement, and perhaps even more to the amazement of those who knew me and were aware of my sentiments, I went along with Watcyn to Carmarthen and found myself for the only time in my life sitting round the table with some of the hierarchy of The Onion, including the President himself, no less. Let me say at once that he seemed to grasp very quickly the point Watcyn was trying to make, and was particularly impressed with his argument concerning the mortality in calves, whilst the reception we received was friendly and courteous.

I was impressed, too, by their Assistant Secretary, Miss Mary Jarvis, and by her letters afterwards which showed that every effort was being made to let light into dark places. But you don't beat 'the system' that easily, and nothing came of it.

In one letter which she received from the Secretary of the M.M.B. it said, 'Our Technical Director, Frank Harding, covered this point in an open reply to Mr Watkin-Richards in the *Milk Producer* and I attach a copy of this open letter in which you can see Frank Harding has used a number of calculations to show the point he has made.'

All perfectly true, of course, but it said nothing about Watcyn's letter which the *Milk Producer* had not used, nor about the intra-muscular injections. Still, they had elevated Watcyn to the distinction of a hyphen, even if the spelling was not quite right.

As a gesture of good-will, and to help in a worthy cause, Watcyn had agreed to co-operate with the M.M.B. in their trial testing scheme, but the Trades and Standards people refused his request to test his milk to see whether it was fit to put in the tank.

I am not sure whether Watcyn is an eternal optimist, or just a glutton for punishment, but, by way of keeping his options open, he also tried the Divisional Health Officer of the Local Health Authority, having forgotten presumably how useless they had been in the case of the river pollution as related in the first chapter. The reply said, 'I have been in contact with our Public Health Analyst and he explained that it is difficult to quantify a level of antibiotic in milk which would render it unfit for human consumption. The levels in excess of this figure do not necessarily mean that the product is unmarketable.'

Helpful. And there was more in the same vein.

The Welsh Consumer Council also turned out to be another case of 'sending the fool further.' How many such characters do the producers of life have to support?

One of the letters received by Mary Jarvis was from the Secretary of State for Wales, in the course of which he said:

It may well be, as you say, that some milk producers, using some intra-mammary preparations, having followed the manufacturer's instructions on dosing and withholding periods, could nevertheless, find themselves in a position where their milk supplies have failed the MMB tests.

I am advised however that this is not a widespread problem and it can be minimised by dairy farmers taking veterinary advice, since there are many preparations available which should ensure that the more stringent current requirements are met.

Helpful again. And here, once again, was also Watcyn's own M.P. (Nicholas Edwards), so there was no point in going to him.

Watcyn did, however, try his Euro M.P., Ann Clwyd, and, when she moved on to other pastures, the case was taken over by her successor, David Morris.

And at this stage I'm feeling a bit like Huckleberry Finn again, and if I'd knowed what a trouble it was to make a book I wouldn't a tackled it. Suffice it to say that, though it dragged on with the Euro M.P. for another three years, nothing came of it.

During this time, however, a well-wisher 'on the inside' sent Watcyn a report by the B.E.U.C., a consortium of consumer organisations in member states of the European Community, which had been received by the Trades and Standards Department of Dyfed County Council in January 1983. Many parts of it tended to support the points Watcyn had been trying to make and, towards the end, it said:

Controls over manufacture are necessary to ensure the safety and efficacy of antibiotics. Manufacturers should be required to prove by thorough tests that their products meet these criteria. The evidence should then be carefully examined and checked by an investigative public authority with facilities for carrying out independent research. Only if this authority is satisfied may a licence be granted for a fixed period with review at the end of this time or if new evidence becomes available.

Current licensing procedures operating in many countries need to be improved. In many cases they rely exclusively on a written dossier submitted by the manufacturers. Backlogs of applications build up and the responsible authorities become too passive, tending to approve products without any in-depth evaluation or independent testing.

Such is the position in the United Kingdom.

Even though others fell by the wayside, Watcyn, nothing daunted, pressed on. Two letters he wrote during those few years are perhaps worth recording. The first was in April 1985 to that lively agricultural paper, *Farming News*.

I was incensed, rather than merely interested to read your brief note (April 5th) to the effect that the drug firms and the N.F.U. are concerned that the Government plans legislation to standardise withdrawal periods for livestock drugs by Dec 31st.

The drug firms have grown fat enough over the years to be well able to withstand the loss of whatever drugs on their shelves at the moment have to be withdrawn. But the fact that the N.F.U. are concerned defies comprehension. It was because of the N.F.U.'s purblind attitude over this serious question, and when Sir Richard Butler refused to meet me over another issue, even though he was coming to the county, that I cancelled my membership of that organisation.

The only encouraging thought is the fact that the

Government seems to be intent on taking long overdue action at last in line with world health standards.

Later that same year Watcyn noticed a pronouncement by Mrs Thatcher on the question of drugs, so he thought the occasion could be apposite to drop her a line by way of encouragement, which he did as follows:

Dear Mrs Thatcher, I am delighted that you are taking an active interest in trying to stamp out the abuse of drugs in our society. Whilst you are at it, could you take an interest in the abuse of drugs in the farming industry which is being carried out for the benefit of drug manufacturers, with the connivance of the Milk Marketing Board and your own Ministry of Agriculture?

I refer specifically to the scheme which is due to start on Jan 1st whereby the sensitivity of the test for antibiotics in milk will be doubled. I have no objection to this except that milk producers are being hopelessly misled by the instructions given for the duration of the period for which contaminated milk should be withheld.

I cannot approach my own M.P. as he happens to be Mr Nicholas Edwards, and I am hoping to be taken to court by him fairly soon in his capacity as Secretary of State for Wales.

If and when that happens I hope that much useful national publicity may result and be of some help in the campaign I have for too long been having to wage on my own in this field. I am writing to you in the faint hope that as a parent you will show some concern and that, as a person who seems to have more sincerity than the average politician, you may have a more realistic grasp of what is being done than so many of the idiots it has been my misfortune to have to deal with so far.

Now, wasn't that a nice letter? And surprise, surprise! Watcyn had a reply from somebody at the Min. of Ag. and Fish.

All right, then. You try beating 'the system' some time, especially if you happen to be the only one in step, and let me know how you get on.

At least he found out the hard way over another issue that it would have been a waste of time trying to take any part of the affair to the Ombudsman, knowing that that outfit is just about as strong in the parasitical stakes as all the other parasites into whose antics it is supposed to be their function to enquire.

Watcyn's flirtation with The Onion did not last long. In due course a little dicky-bird passed the word on to him that things were not going to go his way because one of the head sherangs of The Onion had said in a Trades and Standards committee meeting that he would take care of Watcyn. The little dicky-bird in this case was one of the better types who had been helpful to Watcyn, as far as it lay within his limited powers, and he was one of those who took early retirement, having become sickened of the whole business.

As far as this particular head sherang of The Onion was concerned I can do no more than recall the words of the late Rees Owen of Camrose, who gave his life to the service of his fellow farmers through the N.F.U., 'It's unfortunate, but a fact of life nevertheless. When a farmer becomes a councillor he usually forgets that he's a farmer.' Who was it said, 'Man changes but little, God never.'

So Watcyn was still up against some of his fellow farmers. What was that about another sheep?

But, when he cancelled his brief membership of The Onion, there was something far more sinister behind it than anything which has so far been related, and the telling of it will take its proper place in due course.

Indeed, even as I write these lines, the story is not finished, and I do not know how the book will end.

Huckleberry Finn just didn't know nothin'.

CHAPTER SIX
The affair of the tractor safety cab

Although it was over the antibiotics that Watcyn went to The Onion, he had cancelled his membership of the N.F.U. because of their failure to support him over the saga of his tractor, on top of their earlier failure over the antibiotics. And it was over the business of the tractor that he had found out the hard way, like many disillusioned souls before him, what a complete and utter waste of time it was to go to the Ombudsman. It was, I believe, a proud claim of Little Harold, that the creation of this Office was one of the great achievements of the Wilson administration. So be it. And why need Britain tremble?

So, having finished with the antibiotics, and having already made passing reference to the tractor, it is time to go into more detail of that unbelievable affair.

Some time in the 1970's Watcyn bought a Steyr tractor from a local firm which subsequently went bankrupt. He reckoned then, and he is still of the same opinion, that the Steyr tractor is one of the best in the world. Various national and world trials and the winning of many championships would seem to confirm this opinion.

It was but natural, therefore, when the time came to think in terms of a replacement, that his mind should turn once again to another Steyr, and he saw the tractor he fancied on display at the Pembrokeshire County Show in

August 1978 on the stand of a local dealer. He was told, however, that he could not buy that particular tractor, as the safety cab, which was the type supplied with the tractor in its country of origin, did not conform to British rules and regulations. Something about two decibels. Even so, decibels or not, the modern tractors are nearly all fitted with radios, and the cacophony of the jungle can be belting out like glory be for as long and as loud as you like. No wonder the tractor drivers have to wear ear muffs, when they ought to be listening for the sound of the engine in relation to the implements with which they are working.

For reasons which are irrelevant here Watcyn eventually bought the new tractor from the Steyr importers, Bridgemans of Newbury in Berkshire. The deal was struck in March of 1979 and Watcyn was allowed £5,523 on his old Steyr 650A against the new Steyr 768A costing £11,523. A pick-up hitch was fitted as an extra option at £70. The gear-box was stated to be fully synchromesh and the invoice, when it eventually arrived, confirmed this. In the fulness of time, as the dispute developed, the solicitors for Bridgemans said in a letter, 'The tractor was not offered as having a full synchro gear-box. The gear-box fitted was as seen and tried by your client when he tested the demonstration model. The invoice was perhaps carelessly drafted.' Knowing what I know of solicitors I would say that this was something of an admission. The fact that Watcyn never had tried out any demonstration model is immaterial, and we can forget about the gear-box as well.

More importantly, the cab was described as a 'Luxury Cab.' As Watcyn said, whether it was luxury or de luxe, at £1,500 it should have been gold-plated. So we can now forget about the tractor, as well as the gear-box, and concentrate all our thoughts on the cab, for that was the cause of all the trouble that was in the offing.

Having done the deal, Bridgemans collected Watcyn's old Steyr tractor and said that the new Steyr 768A would be delivered in a week or ten days. That was early in March.

Time went by and, in the absence of the new tractor, Bridgemans lent Watcyn a tractor to be going on with, and this was eventually referred to as a demonstration tractor. What it demonstrated I am not too sure unless it was that the whole thing was a complete cock-up. This is not an expression I was brought up to use in the best society but, although I am richly blessed by not having a television set, I understand that it is an expression which is frequently heard on the idiot's lantern, so I suppose it must be all right.

Watcyn therefore asked for his own tractor back but was told that this had been sold and that his new tractor would arrive in a week's time. It did not, so they sent another so-called demonstrator which was also unusable. By this time Watcyn was busy on the silage and having to borrow tractors from wherever he could.

Then, in August, the new tractor arrived and, before it ever came off the lorry, it could be seen that all was not well. No bill, no invoice, no number plates, no registration documents and no pre-service manual filled in. Yet Watcyn could not send the tractor back, otherwise he would have had no tractor at all.

So now, all these points apart, we come to the crux of the matter. The trouble had all been caused by the new cab which seems to have answered the requirements about decibels and that was all. It had been designed and built by a firm called Retford Sheet Metal Co, henceforth to be known as R.S M., under the benign guidance of the Health and Safety Executive, henceforth the H.S.E.

Whereas, on the face of it, it might be thought that a

safety cab should be designed to fit a tractor, the idea now seemed to be that the tractor should be rebuilt to fit the cab.

Watcyn complained, Bridgemans now offered to let him have his old tractor back, and Watcyn refused, saying it was now up to Bridgemans to put the new tractor right. With the wisdom which cometh with hindsight it is all too easy to say that this was perhaps a mistake.

Many moons later Watcyn had to call in various people in support of his argument. Although it is going ahead of the story, it could be helpful to quote their reports at this juncture so that the position will be clear from the outset.

The first report was in November 1979 from the old established firm of Consulting Engineers, W.D.Davies & Co(South Wales) Ltd, who carried out their inspection on Nov 2nd. Their inspector's report said:

I understand a part exchange deal for this tractor was arranged in March 1979 with Messrs W.R. Bridgeman & Son Ltd, Speen Lane, Newbury, and delivery was promised in three weeks from this date.

The tractor was delivered to Mr Richards' address on August 5th 1979. Some control pedals were apparently re-set and electrical faults rectified at this time but the tractor did not have a pre-delivery check carried out.

The tractor bears registration plates number CRD 581V but it has since been established by Mr Richards that this registration number has been already allocated to another tractor of different make and type.

I understand from Mr Richards that a Safety Officer, possibly from the Ministry of Agriculture, has inspected the tractor on more than one occasion and taken photographs. A report from this source on their findings would be most useful.' (Yes, indeed, it would have been most

useful, if they'd had the will or wit to find anything – R.H.)
The chassis number from the plate fitted to this tractor is 760A–29505–3065 whereas the badge fitted to the bonnet states 768. It is yet to be established what actual model this tractor is and the gross horse power.

Most of these points were sorted out more or less satisfactorily eventually although, to the uninitiated, it no doubt sounds a bit like Fred Karno at his most hilarious. It was the business of the safety cab, however, which was the gravamen of Watcyn's complaint and, on this score, the report continued:

The main complaint with this tractor is that the modifications carried out to the tractor to suit this type of cab are not satisfactory.

The cab bears an identity plate quoting RSM/KEP cab model RSM/D10. Approved for use with Steyr 768/768A. 658/658A. Retford Sheet Metal Co. Ltd. Retford, Notts.

We would comment on the modifications to the tractor as follows: –

1. The rear cab mounting bolts prevent the removal of the hitch stabilizer arms, unless the cab mounting bolts and brackets are removed.

2. The stabilizer mounting brackets have been lowered from their original mounting position which prevents a satisfactory operation of the lift, as the mounting brackets are below the level of the lift arms.

3. The right hand stabilizer bracket has fractured on the axle brackets, no doubt caused by the usage of the lift with the position of the stabilizer bracket as stated in item 2.

4. The hitch operating lever has bent in use. It is impossible to operate this lever unless one leans out through the rear window of the cab. It should be possible to operate this lever from within the cab.

5. The independent foot brake pedals have been extended and the lever which locks the foot pedals together across the base of the pedals is not satisfactory, in that it does not remain in the locked position.

6. The control levers for the ground PTO, crawler box and PTO are very difficult to operate and are touching one another.

7. The steering column has been extended and with the engine running a lot of vibration is transferred to the wheel.

8. The windscreen is very difficult to open and close.

9. The driver's seat is unsatisfactory and appears to have no damping action.

10. The hydraulic top link fouls the cab when raised to its highest point.

11. The wiring on the lamps to the rear is stretched to its limit and requires extending and securing clips fitted.

12. The right hand cab door is very difficult to open and close. The rear view mirror hits the cab frame when the door is opened. The mirror has been removed to prevent damage.

13. The left hand cab door appears to have an incorrect glass fitted as there is a large gap in the front corner.

14. The paint finish on this cab is generally poor and unless improved rusting will occur.

To conclude, the modifications carried out to this tractor to suit this type of cab are far from satisfactory. The cab may well be an approved cab but we doubt whether the modifications are up to the approved standard.

I would suggest a report be obtained from the Safety Officer who inspected this tractor.

To put it another way, at the behest of the bureaucrats, one of the finest tractors in the world had been reduced to an abortion and an abomination. And we can return to the Safety Officer and any others of similar ilk in due course.

A month after the report, another inspection was carried out by the Honorary Organiser (Agricultural) for the Royal Society for the Prevention of Accidents, Mr Chambers B.Sc., B.Agr., F.I.Ag.E.

As Max Boyce said in one of his classic stories when referring to the attitude of an aggrieved member of an unsuccessful choir from North Wales at a particular eisteddfod, 'The adjudicator was from down southed – from Carmarthen. Madame Clara Teify Jenkins – F.R.A.M., L.R.A.M., F.R.C.M., Ph.D.Mus., Mus D. (Oxon) – what do she know about blutty music!'

Well, there we are. In a similar way maybe the bureaucrats were not impressed with Mr Chambers' qualifications and experience.

His lengthy report went into even more detail, and painted an even blacker picture, than that of Messrs W. D. Davies and Son.

A year later a third independent observer was called in and he was Mr Malcolm James who had, in fact, been the sales representative involved when Watcyn had bought his first Steyr tractor. I don't know how much he did or did not know about 'blutty music', but he was an eminently practical man.

Having outlined his background experience of agricultural machinery, and, bearing in mind that, as a result of demonstrations, he had sold twenty-three tractors (Steyrs) in his first year, he was in a position to make informed comment.

His report was a repetition, or confirmation, of the

other two, plus elaboration on a few more faults. Whilst it is evident that the job was a complete bodge, and that so many aspects were unsatisfactory, it is the question of the cab and the safety aspects with which we are concerned. It was unfortunate that, because of the bodge, it was all too easy for those who should have been concerned with safety to ignore that aspect and use the bodge as a red-herring to make it appear as if Watcyn should have been looking elsewhere for redress.

As we have seen, when the new tractor eventually arrived and was found to be so unsatisfactory, Bridgemans offered Watcyn his old tractor back. That was the one which they had previously said he could not have back because they had sold it. This took place when the principle of the firm, Mr John Bridgeman, came down to Bunkers Hill, with the sales representative, Mr Malcolm Broughton, to try to rectify the various faults. The latter did, indeed, do his best and, although he was the sales representative, came back himself three weeks later to do some of the servicing which so far Bridgemans had failed to do. But when Watcyn asked him what was to be done about the cab, he was told that that was the responsibility of R.S.M. It was at this stage that Watcyn, in his blind and blushing innocence, poor simple fool that he was, had the bright idea of going to the Health and Safety Executive. It is a sobering thought, is it not, when we realise how many parasites the producers of this country have to carry on their over-taxed backs, but carry them they certainly have to.

A farmer friend of mine was complaining to me the other day that, at sixty-eight years of age, he was still out milking his fifty cows at six o'clock in the morning, and the result was that the tax man took most of his old age pension off him. If he sold his bit of a farm they would clobber him for Capital Gains tax and he couldn't live on his pension. Not that he objected to paying his fair share of tax, but he objected to what was happening to the money. Every day he could see the local government hordes travelling along the main road past his farm, going to their offices long after he'd finished milking in the morning, and driving home again in the afternoons long before he would be due to start milking in the evening, all driving their latest registration cars. 'And the thing is,' he said, 'the buggers are laughing about us.' And, he could have added if he'd only thought of it, in a land where Mrs Thatcher was still paying lip service to individual enterprise, long live the small business man.

It is another sobering thought that so many of these parasites are ever ready to jump on, and interfere with, the individual, especially if he is in business. But ask them to do something useful to protect the interests of these same individuals and see what you get for your trouble.

Anyway, Watcyn tried the H.S.E., and one of their assistants called to see what all the fuss was about. Then he went back to his cosy office and wrote to Watcyn to say he had discussed it with his senior officer and that they had decided it would be 'extremely difficult to decide which items, if any are attributable to design faults in the case of a tractor that is not new.'

So, a tractor delivered on August 5th and seen by this idiot on August 29th, is not new. By way of a further gratuitous insult he suggested that Watcyn should complain to his supplier or possibly 'take up the matter with the Office of Fair Trading.' Helpful again.

Shortly afterwards Watcyn had another visit from another little creep. Oh, yes, there are plenty of them. And, as a result of what happened, I am reminded of a situation in my own farming life when this little character used to come round on behalf of the Min. of Ag. and Fish. to enquire into the business of acreage payments. It was a job he did in tandem with his other function of sneaking about to see what he could find by way of a safety guard missing or maybe no handrail on the steps going up to the loft. A nasty bit of work he was. Then I heard of an incident concerning a visit he paid to a farm for acreage payment inspection, and he saw the farmer driving out to the field and about to hitch up to an inadequately guarded implement. So this little creep hid behind the hedge, waited for the farmer to commit the heinous offence, and then caught him red-handed. However, the legal characters at the Min. of Ag. and Fish. were not impressed. They did not think he was the sort of witness they would be pleased to call and so they did not prosecute. Maybe under cross-examination he would not have created a particularly good image for the Ministry. And 'Peeping Tommy' had 'a black against him' as these characters say, so that he never would get much by way of promotion. But then again, he wouldn't be sacked, and he could always be moved sideways and be sculling around with his sly and sneaky ways to continue as a nuisance and a nasty bit of work to boot.

As a result of this visit, Watcyn then had another visit from the original H.S.E. assistant inspector who wrote to him five days later. He made not one mention of the Steyr tractor with its twenty-odd cab-induced faults, but drew attention to two minor safety faults on a 1961 Fordson Dexta tractor which had been standing alongside it.

And he also, with a touch of inspired genius, drew attention to a minor fault on an elevator which had been nowhere near Watcyn's farm at the time of his visit because it had been on loan to a neighbour. From which it is reasonable to assume that it was 'Peeping Tommy' who had seen it on the occasion of his visit and passed on the word.

Not one of these characters had been prepared to accept Watcyn's invitation to drive the tractor in order to test it, but had at least been honest enough to admit that they didn't know how to. And yet, in due season, the Ombudsman had the gall to talk about inspectors' reports. Worse still, there never was a report by the H.S.E. They had approved the cab in the first place and Watcyn Richards of Bunkers Hill, in the parish of Camrose, was no more than a squalid nuisance.

Much was also made of the fact that Watcyn had not paid for the tractor during the time that all this was going on, but there was no mention of the fact that no invoice was sent until Nov 1st.

It was following this that Watcyn had the first of the independent reports to which reference has already been made. Then Bridgemans agreed to fetch the tractor to carry out the necessary remedial works on their premises at Newbury. Before the tractor went, however, Watcyn had the inspection of the tractor on his farm by Mr Chambers of Ro.S.P.A., and Bridgemans lent him a tractor to be going on with. They seem to have been quite good like that, although it was not long before their solicitors were writing to say their clients wanted it back because it had been sold.

A month later Watcyn went up to Newbury to see how matters were progressing and there was something of a contretemps, if that is the right word, and it's probably as good as any other. Pronounced kong-tr-tong, so teacher used to tell us in school, but we're living in changed and changing times. And whichever way they pronounce it these days, you shall judge for yourselves as to whether it was indeed a contretemps or whatever, as the saying is.

Watcyn had had a lift up to Newbury with Mr Malcolm Broughton, who was still being as helpful as possible, and the idea was that Mr Chambers would also be there to meet them.

Unfortunately, he had trouble with his car on the way, and so he was late, but it will be readily understood from what has so far been written that there would have been no shortage of topics for disputatious conversation and discussion. To his disappointment but not, perhaps, to his surprise, Watcyn found that the tractor had not been touched.

As the well-known phrase has it to be, one word followed another, and the upshot of it all was that Mr Bridgeman called Watcyn a Welsh bastard.

My function is merely to record the facts, and it would perhaps be out of place for me to express too many personal opinions. So it is not for me to say whether he should have consulted the Race Relations Board on this issue, although I have no doubt there are many law-abiding citizens who would have urged this course upon him. On the other hand, if that organisation, which is one of the increasingly popular growth industries destined to put the Great back into Britain, and let me hasten to stress that I have no personal experience of them, were to prove to be no more useful than all the other useless clutch or gaggle of non-collaborators we have met in these pages, then it would have been just another fatuous waste of time. In any case, I honestly believe that we can forget any question of Race Relations Board involvement, and that any discussion on such a possibility would have to be purely hypothetical, academic and irrelevant, because as Mr Bridgeman mouthed the obscenity to the effect that Watcyn was a Welsh bastard, he was also concomitantly charging at him swinging a big iron spanner.

Let us not dwell on the unhappy incident. Watcyn led with a left to the guts and dropped him with a right cross. Then some characters came and carried him into the office – Mr Bridgeman, not Watcyn.

These days, so many years later, Watcyn is a little hazy in his memory as to the exact sequence of events which followed immediately upon this particular phase of the confrontation. Perhaps I should quote his own words on the matter. 'All I can remember,' he said, 'was that I heard a hell of a crash and I looked up and there was this chap – a mechanic I think he was – coming flying out backwards through this glass door with his arms flung up in the air like something out of Starsky and Hutch. And I thought "What the hell's going on here then!" It was real weird.'

Finally he decided there must have been another dissatisfied customer in there.

Perhaps I should explain, for the benefit of the uninitiated, that Starsky and Hutch, as I understand the term, is derived from some programme or other on the one-eyed monster which is transmitted for the delectation of the younger generation. But I have no first-hand experience of this, so I would prefer not to be quoted.

It should not be inferred from this that Watcyn is of a pugilistic nature. Whilst the incidents to which it has been necessary to refer in these pages would not have shown Watcyn to be, in many ways, a rather gentle character, his own philosophy is to walk away from physical confrontation as far as possible. But his father was a bruiser of some repute and his saying was always, 'Un yn bon gern gyntaf a gofyn cwestion'e ar ol.' (One in the jaw first and ask questions after). Presumably Watcyn had learned one or two things from him and, in this moment of extremity, old habits died hard.

The scarf will never know how lucky he was! Watcyn is neither a moral nor a physical coward. Nor should he be regarded as fair game for bullies and fools. Shortly after that Mr Chambers arrived and they were provided with sandwiches and a pot of tea, and it is a delight to be able to record an episode so very friendly and civilised.

Following this visit the tractor was returned to Watcyn and, foolishly as he subsequently realised, he agreed to accept it as satisfactory, with the suppliers agreeing to pay all his legal costs and other costs involved. How well advised he was by his then solicitors it is not for me to say. And, if the wiseacres choose to criticise him on this score, let it be remembered that there was, as we have seen, so much else happening in his affairs at that time, and he was under tremendous pressure, to say nothing of the strain it was all imposing on his wife and family. And R.S.M. had gone bankrupt anyway.

In spite of all this, however, he saw no reason why the Ombudsman should not be called upon to examine the antics of the H.S.E. and its assorted minions, and he wrote to his M.P., still incidentally Secretary of State for Wales, on Feb 28th, 1980, to this effect.

I do not propose to dwell on too many wearisome details but, to give those who have no experience of such matters some rough idea of what they get for their money as taxpayers, let me just mention that, over a year later, on March 16th 1981, Watcyn had a visit from two characters from the Ombudsman's office. Ever so charming, but no, or had you guessed, they didn't wish to try driving the tractor. Indeed, one of them admitted, it was the nearest he had ever been to a tractor in his life.

In September of that year – the same year, fair play – the Ombudsman issued his report. It ran to fourteen pages and, publishing costs being what they are, I have no intention of quoting it in full, especially as it is a complete and utter travesty.

Although they would strongly recommend a manufacturer or supplier to undertake appropriate corrective action in respect of products already supplied, H.S.E. have no powers under current safety legislation which would enable them to undertake the correction of items on a machine which has already been sold.

Only powers, apparently, for the launching of a death trap onto the market.

One of H.S.E.'s assistant agricultural inspectors went to see Mr Richards on 29 August 1979. H.S.E. files contain no details of the items discussed during the visit and the agricultural inspector cannot now remember much of what happened. However, he told my officers that he had listened to Mr Richards' complaints about the tractor and cab and had reported back to his senior officer.

This was the joker who had referred, four weeks after delivery, to a tractor 'that was not new'.

Accidents were known to have occurred due to the operator's failure to link the pedals together before going on the road, but, so far, there had been no evidence to suggest that vibration had been the cause of this link disengaging.

Evidence is it? How long is it then since dead men have been able to testify?

H.S.E. tell me that it was unusual for a farmer to make a complaint about the safety of a machine. It was more usual for H.S.E. to raise these matters during a routine safety inspection.

Yes, of course. Even to raising matters in the case of a machine which was not on the farm at the time of the inspector's visit.

Mr Richards received three visits from H.S.E. inspectors, two of which were directly related to his complaint about his tractor. The inspector who undertook the first visit now accepts that it was an unfortunate piece of drafting on his part to have referred to Mr Richards' tractor as 'not new'. I agree that this can have done little to inspire confidence in H.S.E.'s handling of Mr Richards complaint up to that point. After the second visit a full and comprehensive report was made by the senior inspector.

The only full report that was made was on the eighteen-year-old Fordson Dexta with two minor faults, and the bale elevator which had not been on the farm at the time of his visit but had been spotted previously by 'Peeping Tommy'.

On the evidence of my investigation I feel that it would have been more in keeping with the spirit of the 1974 Health and Safety at Work etc. Act if H.S.E. had made somewhat more effort to reach agreement with Mr Richards on what needed to be done to correct the safety-related defects about which he had complained. When I showed H.S.E. the two independent reports which had been commissioned by Mr Richards they agreed that if they had been shown them earlier they would probably have raised the brake linkage problem again with the suppliers (who had previously told H.S.E. that they would correct it). They would also have regarded the gap below the automatic hitch hook and the associated guard plate as potentially dangerous and a fault to be remedied (a fault of the tractor not previously brought to light and having nothing to do with the cab fitting). Nevertheless they remained of the opinion that many of the points in the reports concerned matters of quality and were therefore outside their strict ambit.

There is a short answer to all this, but I have used the expression 'Cobblers' a couple of times previously and too frequent repetition is not good style.

Then, in the final sentence, the Ombudsman says, *'After thorough investigation, I am unable to uphold.'*

Well, after one year and seven months, thorough is not necessarily the first word that springs to mind.

Much of what happened afterwards has already been touched upon in one way or another. Watcyn failed to elicit the support of the N.F.U. and tried The Onion. Theirs was the classic example of the Welsh idiom 'tan shafins', which means fire from wood shavings. Very expressive really, because nothing much comes of it and it doesn't last long.

By 1983 R.S.M. had already gone out of business and Steyr had withdrawn their concession from Bridgemans and no longer had dealers in the United Kingdom. I don't suppose for one moment that that had any more to do with Watcyn's business than the revision of the withholding period for antibiotics by the drug firm, or the change in advice by the M.M.B. to the effect that contaminated milk could be fed to livestock.

But take heart, all you taxpayers. The H.S.E are still in business, and you hear from them whenever there is a national disaster, such as a fire on the London Underground, or when a stand collapses at a big football match. It is always good to know at such times that the nation's welfare is in safe hands, and there is a feeling that their pronouncements must be of immense comfort to the bereaved and to any survivors.

CHAPTER SEVEN
Tests don't come easy

You can't win 'em all. I am not sure whether that is a cliché or something Confucius said, but it happens to be a fact of life. At which point we can return to the subject of antibiotics and so much which came in the wake thereof. Watcyn put up a good fight, the gentlemen of the media had a field day, and a good time was had by one and all.

When we last referred to the unhappy subject Watcyn was writing to Mrs Thatcher. He had long since become disenchanted both with the N.F.U. and The Onion, and he had found out the hard way, over the affair of the tractor cab, that to have recourse to the Ombudsman was about as rewarding as looking for a pork chop in a synagogue. So he decided to go it alone, as the saying is.

What he wanted was publicity. Not for himself, but for the cause which he had espoused. And he certainly had it. Whether it was any good to him or not is a moot point, for in the final reckoning he did not have the publicity he wanted or when he wanted it, as I must now explain.

Having tried every channel he could think of, and having been disillusioned with them all, he embarked on a campaign of non-cooperation. This did, admittedly, involve a certain degree of obstreperousness in certain instances, but the basic principle was really one of non-cooperation. Is it not amazing how civil servants and such think they can turn up on your farm as though by divine right and expect to be attended to? Some of them, especially the smaller fry, take a very poor view when they are told to go away and come back some other day by appointment. Would you call that unreasonable? After all is said and done, we have to make appointments if we are ever sufficiently at a loose-end or sufficiently maladjusted as to want to go and see them. So why should not the same principle apply in reverse?

In the autumn of 1982 Watcyn had a notification from his own vets to say that they were coming to carry out the routine tests of his herd for brucellosis and tuberculosis. They would, in this case, be acting on behalf of the Min. of Ag. and Fish. and it was normal practice. To notify him was a formality, just as it was a formality for him to tell them not to waste their time. He also confirmed this in writing to the Ministry.

Following that the Divisional Veterinary Officer took over and said he would come to do the job himself. Watcyn had told him there was no point in coming and he would not cooperate until such time as the Ministry agreed to a meeting round the table with himself, the M.M.B., the Chief Trading Standards Officer, the Public Analyst and anybody else whose observations could be of use to a grateful nation. Notwithstanding this, the gentleman came to Watcyn's farm which is at the bottom of a narrow lane about four hundred yards from the road. I don't think I mentioned that. But I have mentioned that Watcyn was being obstreperous. The Divisional Veterinary Officer had to reverse his car the four hundred yards back up the lane.

Before indulging in this manoeuvre he intimated to Watcyn that he would come again another day, which he did, in March 1983, accompanied by colleagues, and supported by the police, with press men and photographers various bringing up the rear.

The police explained to Watcyn that, in certain circumstances, they had the power to arrest him, so he said he was a law-abiding citizen and in that case the Divisional Veterinary Officer had better drive his car into the field out of the way to make room for some of the others, which he did. And then Watcyn parked his tractor in the gateway and said what about this meeting that he was on about and, whilst they were at it, wouldn't it be a good idea if the meeting could be public so that the world could know what was happening?

So the police said they would move the tractor themselves and that was about the only laugh Watcyn had all day, because you know, and I know, as Watcyn knew only too well, which tractor it was, and he told them that if they were willing to try driving it they were better men than those who had come to see it from the Health and Safety Executive, the Ombudsman and all the rest of them. So the Divisional Veterinary Officer, at Watcyn's invitation, used Watcyn's telephone to have word with the Welsh Office's Agricultural Department at Aberystwyth to ask them what about this meeting, and they told him never mind about any meeting but go and carry out the test. Easy for them, of course. So he told Watcyn he was going to carry out the test and Watcyn said that was good and he would watch it with great interest.

Even with the full backing of the Dyfed Powys constabulary the man knew it would be far beyond his authoritarian powers to round up and pen the whole herd for testing, but give him full marks for having more sense than to try. They settled in the end for testing a dozen young cattle which they found in one shed. Watcyn thought little of the way the job was being done, was horrified at the blood about the place and asked the police to intervene. He thought that as they were there anyway, and as he and his property had as much right to their protection as those who had ensured their presence, then they might just as well make themselves useful.

The testing was abandoned at this point with Watcyn intimating that if they ever came again then the R.S.P.C.A. would have to be there as well. This was a new thought, but there is no record of his having suggested their presence at any round-table conference, and with all that we hear to the detriment of the R.S.P.C.A. these days maybe it was just as well.

It is perhaps worth noting that, on the evidence of the result of the test on these ten unfortunate animals, Watcyn subsequently received a notification from the Ministry that his whole herd had passed the test. Success to temperance and three cheers for the working class.

As a preliminary skirmish this was no more than promising. Subsequent Ministerial blandishments having been consistently resisted, the balloon really went up a year later in February 1984.

Normally, in these cases, the test is carried out comfortably by the vet with the assistance of the farmer. That is not good enough for officialdom. On this occasion it required four contractors hired by the Ministry, plus two Ministry vets and their two assistants, whilst four police cars disgorged a sufficient number of members of the force to gladden the heart of any film producer. Expense no object. There were also various characters from the legions of Dyfed County Council and Preseli District Council. A police car headed the convoy down the lane, followed by the contractors' lorry, followed by a police panda car, followed by the marching cohorts of the index-linked pension brigade, whilst the gentlemen of the media scuttled here and there and every which way in an endeavour to secure the best vantage points. The term 'gentlemen of the media' also includes television crews.

By some chance the Steyr tractor was parked in the gateway, but nobody had been sufficiently foolish as to bring a car down to the farmyard this time. The usual discussion ensued, the tractor was moved and the test, for what it was worth, took place. The term 'for what it was worth' is used because, out of more than one hundred and twenty cattle on the farm at the time, only fifty were tested. According to the rules and regulations all adult cattle must be tested. In this case no fewer than thirty adult animals were missed.

Following all this the account was rendered in the sum of £350 plus Value Added Tax which brought the total to £402.50. The bit about the added value is always hard to stomach. Who benefited from the added value in this case it is difficult to say, unless it would have been the gentlemen of the press.

They had full value with their headlines of 'The Battle of Bunkers Hill' and such like.

Watcyn, of course, declined to pay. However misguided others may consider him to have been, according to his own lights he had had good grounds for refusing to cooperate with the Ministry. In addition to his basic grievance, he then argued that he had not engaged the contractors and that, in any case, because of the failure to test all the cattle, the test was worthless.

By July the account had been passed to the Debt Recovery section of the Ministry.

Shortly before the massive visitation at the end of

February, Watcyn had received a telephone call from the D.V.O. to say that his Regional Officer, Mr Butler, had agreed that the time had come when the Ministry must be prepared to sit down round the table and talk about it. At the same time the M.M.B. were arranging for a meeting to be held in Llanelli and this was fixed for May 10th.

In many ways a trusting soul, Watcyn had genuinely believed that the routine test, which was to cause so much fuss, would have been deferred until such time as, what he believed to be the promised meeting, had taken place. Likewise, he had believed that it would be just the one meeting between all the interested parties. On both counts he was wrong.

As we have seen, the test was carried out. And, by the end of April, Watcyn was writing to the Assistant Secretary of the Agriculture Department of the Welsh Office in Cardiff, in reply to an invitation etc. etc:

What makes you think I would derive any satisfaction from paying my own expenses to come all the way to Cardiff just to talk to you?

A meeting has been arranged for May 10th, at 11.0 a.m., at the M.M.B.'s regional office at Llanelli, between the M.M.B.'s technical people from Thames Ditton, the F.U.W., Trades and Standards Dept., and myself. That is where a Ministry presence could be of some help, and I am now asking you to arrange for this.

The Min. of Ag. and Fish. would have had no intention of being represented and, at the specific request of the M.M.B., the Chief of the Trades and Standards Department stayed away. A week later, however, he produced a report following a survey organised in conjunction with the Welsh Office Agriculture Department in which he said:

A certain satisfaction can be drawn from the results

that approximately 83% of all the samples taken showed acceptable levels of antibiotics after four days following calving, but I am concerned with the instances when residues were present beyond this time, one in particular for nine days.

Antibiotic presence in milk is highly undesirable and I view with some concern that the Minister, having announced early in 1982 the reduced limits of antibiotic residue, chose then not to effect the transition until 1st January, 1986. I am advised that this exceptional delay is to enable the drug companies to bring their labelling, data sheets, etc. into line with the new standards, but in the meantime we have a situation where certain drugs are marked which work to a maximum level of 0.05 International Units per millilitre for the withholding period, yet the Milk Marketing Board penalise the producers when the level exceeds 0.02, even to the extent of advising them when it exceeds 0.01.

Maybe it was felt that an official who held such views would not have been much help to the Board and which could have been why the Regional Manager, who was arranging the meeting, asked him to stay away. The Min. of Ag. and Fish. were not represented, so that left the Board's Technical Director, who came down from Thames Ditton, a couple from The Onion, and the Board's then Regional Committee Member, but Watcyn said he wasn't interested in anything he had to say. The Regional Manager pointed out that the Regional Committee Member was there to represent Watcyn's interests as a producer, but Watcyn's reply was that he didn't represent producers but cows. It was just before the introduction of the new scheme of 'one man one vote' to replace the old idea of so many extra votes for every ten cows. Watcyn, had he known, could also have added that the same Regional Committee Member had had his chance to support Watcyn's interests, had he so wished, when support had been so lacking at the last N.F.U. Milk Committee I attended and to which I referred earlier.

The account of £402.50 having been duly rendered, the wheels were now set in motion for the law to take its inexorable course in the collection thereof. But that is not to say that life was without other diversions, or indeed occasional moments of light relief, before the end was writ to that particular chapter of the saga.

A month or so before the abortive meeting in Llanelli, and a couple of months after the rodeo of the so-called test, Watcyn had a ten-day old Friesian cross Charolais calf die. It was suffering from the familiar white scour, which, as we have already seen, had been causing such a high mortality rate amongst calves. Time was when such calves could be treated with antibiotics and be on their feet again within twenty-four hours. But not any longer. Watcyn's vet pointed to the condition of the calf's tongue, which showed, in his opinion, that the calf had built up an immunity to antibiotics, and prophesied that the antibiotics with which he was going to treat the calf would do no good. He was proved right and the calf was dead in twenty-four hours.

I know sound practising vets who said at that time that the question of immunity to antibiotics had been much exaggerated, but it is a fact that nowadays, several years later, other treatment for white scour is much more in use.

Watcyn felt strongly, too, that farmers should have been allowed to use certain veterinary products which were available in other E.E.C. countries. His attitude to the Min. of Ag. and Fish. would not have been enhanced over this issue when he remembered the business of their trying to stop him using M45/20 vaccine and how he had used it anyway and it had proved to be his salvation.

Well, he had this dead calf, worth about £130 – £150 before its sudden demise, and he thought the Min. of Ag. and Fish. should be given the opportunity to examine it in support of his argument. So he established that the assistant D.V.O. would be in his office at a certain time and then took the calf into Haverfordwest and dumped it on the gentleman's desk. If I remember rightly he was the one who had talked about onward transmission in the case of a letter, but there was no talk of onward transmission in the case of the calf. Watcyn had ascertained that if he had dumped the calf outside the office he could have been prosecuted but, apparently, in dumping it on a character's desk, he was within the law. Well, of course, as Mr Bumble said, 'The law is a ass – a idiot.'

The assistant D.V.O. would appear to have been neither impressed nor amused and was quoted as saying, 'This is not a knackers' yard. Please clear the office.' That may sound like stating the obvious, but you will gather that the gentlemen of the press were again much in evidence and were once more having something of a field day. They had, of course, followed Watcyn in. I don't have the goggle box myself, but those who do assured me it came over very well.

As the autumn shades were making way for winter's gloom and slush, the Debt Recovery Section of the Min. of Ag. and Fish. came up with rather a neat idea. Or so they seemed to think. The Ministry owed Watcyn some money in respect of a grant–aided scheme which he had been carrying out, so they collared the £402.50 and sent Watcyn a cheque for the balance.

But under E.E.C. regulations they were not entitled to do that, so Watcyn wrote to them pointing out the error of their ways and they duly arranged for the sum of £402.50 to be put into an onward transmission situation. That is to say, they paid him.

About the same time something else happened. A neighbour of Watcyn's had an outbreak in his herd of Infectious Bovine Rhinotraceitis, commonly known as I.B.R., and Watcyn was helping him to pen the cattle for the vet to inject them.

I.B.R. usually occurs in cattle under conditions of stress and is a form of virus pneumonia, affecting the nose and windpipe, maybe not unlike the common cold. Now it is a well-known fact that there is no known cure for the common cold and it is equally well-known that most people will try almost anything in search of a cure, which reminds me of the story of the man who had almost lived at the doctor's surgery for week after sniffling week hoping for a remedy until, in the end, the doctor said, 'There's only one thing I can think of now and that's to go home, strip off, get into a cold bath and run round the field twice without anything on.'

And this chap said, 'Will that cure my cold?'
And the doctor said, 'No, it won't cure your cold. But it'll give you pneumonia, and I've got a hell of a good cure for that.'

On this occasion the vet was injecting Watcyn's neighbour's animals with terramycin. Sometimes penicillin could be tried and sometimes streptomycin, but all with equally unsatisfactory results.

There was quite a bit of infection about at the time and there were two things which worried Watcyn particularly. One was the use of terramycin in case he had to use it on a large number of his own herd which had already shown signs of an outbreak. Perhaps this is as good a place as any to make some reference to the difficulties of some producers who have to use antibiotics as part of their dry-cow therapy. Those with large herds which are block-calving, that is with all the herd calving within a few weeks of each other, do not experience too much difficulty because any presence of udder tissue in the milk works through fairly quickly. And it should be remembered that tests cannot distinguish between udder tissue and mastitis cell-count. On the other hand, the smaller farmer is more likely to be calving his herd all the year round, so that there is something there all the time.

The other point which was worrying Watcyn was that many references were being made at that time to a viral infection known as Enzootic Bovine Leucosis, known as E.B.L. This is a type of leukaemia for which there is no known cure and it is a notifiable disease involving compulsory slaughter. It was alleged at the time, was generally believed and has not been denied, that the disease was introduced from the U.S.A. with an importation of Holstein bulls by the M.M.B., authorised by the Min. of Ag. and Fish., but not, apparently, adequately supervised. Although the Ministry are ever ready to supervise others, there would not appear to be anybody to supervise the Ministry.

With talk on these matters being current in the farming community, and being more apprehensive than somewhat at the possibility of having to use terramycin on a large number of his herd, Watcyn telephoned the D.V.O. to ask for how long he would have to throw how much milk away. The D.V.O. said that terramycin was not the treatment for I.B.R. Therefore Watcyn telephoned his own vet, who said that was as maybe but that was the only thing he could think of at the moment. And we must remember that far more is known about the subject now than at that time. So Watcyn again telephoned the D.V.O., made reference to the great confusion throughout the land, and no doubt offered an unsolicited opinion on the action, or lack of it, on the part of the Ministry in allowing these bulls into the country to spread disease all over the place. It is reasonable to suppose that they did not exactly see eye-to-eye and, bearing in mind all that which had gone before, it is not hard to believe that harsh words may have been spoken.

That evening Watcyn was just getting ready to milk when a police car arrived with a Detective Sergeant and a Detective Constable carrying a brief-case. Now Watcyn has always got on well with the police and still does. He recognised this particular sergeant, however, as a well-known 'congenital idiot'.

It would be surprising if the police were to attract one hundred percent perfection any more than any other branch of public service. This one, however, was generally recognised to be as dull as a bag of hammers, to use a well-known Pembrokeshire expression, and was, in fact, so thick that even some of the other policemen had been known to notice it. The C.I., as we are using initials so much in this narrative, and in this case it stands for Congenital Idiot, said he wished to talk to Watcyn about threats he had been making to the D.V.O., so Watcyn said in that case he would have to wait until he had finished milking and that took an hour or so. Then Watcyn told him he would have to wait until he had had some supper, and by this time the C.I. was becoming rather worked up.

It was greatly unfortunate that there was nobody else at home and that Watcyn was on his own, so there were no witnesses.

He went into the house, switched off the outside light, made himself some supper and, when he had partaken, switched the outside light on again and bade the C.I. and his side-kick enter. The side-kick spoke no word throughout and his sole function seemed to be to carry the brief-case.

It seemed that the police had received a complaint from the D.V.O. that Watcyn had threatened his life. Or so the C.I. said. And Watcyn found it hard to believe. Then the C.I. said that Watcyn had to appear before a special court in Haverfordwest the following morning and Watcyn asked for some sort of summons or something of that nature. It was not a procedure with which he was familiar. So the C.I. asked the S.K. (Side-Kick) to open the brief-case and then took out a paper from which he appeared to be reading. Taking the not unreasonable view that he was in his own home, Watcyn grabbed the paper, but the C.I. grabbed it back and returned it to the brief-case before Watcyn could read it. Then the C.I. said he was going to arrest Watcyn and Watcyn asked him whose army would be coming to help him. The C.I. hadn't thought of that and he didn't have any reinforcements to hand. What was more, his S.K. didn't look a very likely candidate either. So off they went and that was more or less the end of that.

Watcyn did, of course, send in an official complaint and, as a law-abiding ratepayer, demanded that action be taken 'against the officers concerned and against the clown responsible for wasting their time.'

I understand that statistics show that very few complaints against the police are ever upheld, and that is not much wonder. In this case, which took best part of twelve months to investigate, the 'congenital idiot' and his side-kick denied having been anywhere near Bunkers Hill on the evening in question. The only satisfaction Watcyn had, if it can be called satisfaction, was that he had the sympathy of a fair few policemen who knew the score. I did not hear tell of any of them shedding tears when, immediately the dust had settled, the C.I. took early retirement. Whether or not in response to a pressing invitation from above was never made known. Ill health, they said.

In the meantime Watcyn decided not to treat his herd when the outbreak of I.B.R. occurred and, as with the common cold, it took its own course and worked itself out, so there was no question of inhibitory substances in the milk.

A month or two after this episode Watcyn would not have needed to be paranoid to come to the conclusion that somebody was out to get him and that, somehow or other, they were getting their lines crossed, as the saying is, when there arrived at Bunkers Hill a demand for £402.50. That was no doubt fair enough as far as the Min. of Ag. and Fish. were concerned, but the demand, from one D.T.A. McInnery, was addressed to Watcyn's teenage son, and the worst thing he had ever done in his life was to allow himself to be caught down by the river with a lamp at night, and serve him right. You get fined for doing things like that, even if you don't have a fish. But pollute the river and send the fish floating downstream bellies upwards and you found authority reluctant to act and, as often as not, you got away with it.

Not unreasonably Andrew Nicholas Richards wrote back to Mr McInnery:

I have today received a letter from you which has distressed me greatly. I have never had any dealings with you in my life and, if this is the way you conduct your business, am very glad.

Could you tell me please what sort of organisation you operate and why you should think I owe you any money?

The apologetic reply was profuse and delightful.

During the following summer, whilst the steamroller of State was taking its time over collecting the outstanding debt, Watcyn received a couple of letters from the Chief Executive Officer at the Min. of Ag. and Fish. which called into question an application of his for a grant-aided drainage scheme which had already been previously approved. The C.E.O.'s name was Mr J.C.Alexander, so Watcyn wrote to him saying:

I am in receipt of your letters of June 19th and July 11th. I note that the first was signed by a Miss B.Burrows. I address this letter to you, however, as I have always believed that it was a waste of time to bother with the monkey when it is the organ grinder who chooses the tune. And I do not much care for the tune you are now trying to play.

The various regulations and restrictions you are now seeking to quote do not apply in my case, as my development scheme was applied for in 1982, and it was agreed that I was eligible and that the job could and would be undertaken in two stages. If this agreement had not been forthcoming at the time I would not have tackled the job.

I do not propose to enter into more detail since it is perfectly clear that an attempt is being made to discover any discrepancy in my files which you can use to victimise me further in my fight with the Ministry and the litigation which is now pending. It would seem that someone has produced this feeble straw at which you can clutch.

I am sending a copy of this correspondence to the Registrar.

Due to the system, or lack of it, which seems to obtain amongst so many of these organisations, there was yet another letter on the same subject on the same date from yet another signatory.

Watcyn wrote to the C.E.O. as follows:

I notice that your letter of July 10th is signed by a different monkey again but the organ grinder remains the same.

What gives you to understand that I wish the claim submitted on Nov 13th 1984 in respect of Drainage Work to be withdrawn?

I have never expressed any such wish. I did say to the field officer that there seemed little point in proceeding if some of your mafia decided to intercept payment on other work on which I am engaged and for which grant is payable, the same as you tried to do once before.

I have already returned the original documents and would be obliged if you would ignore your own letter and proceed with my claim.

I am also sending a copy of this letter to the Registrar dealing with the litigation involving the Ministry and myself.

The scheme was completed and Watcyn was duly paid.

The only other diversion hereabouts was just before the case was due to be heard and Watcyn received a document from the Court to the effect that he was being sued for payment by the Secretary of State for Wales. That was why he had said in his letter to Mrs Thatcher, as referred to earlier, that he was hoping to be taken to Court fairly soon by his own M.P. in his capacity of Secretary of State for Wales.

Watcyn had also found an apparent reluctance on the part of the Ministry to produce the various documents relevant to their case, so he wrote to the Clerk of the Court:

I cannot see my way clear to sign a Certificate of Readiness until the Ministry agree to all the relevant documents being produced. Either that or they engage contrac–

tors to drag me before you the same as they did with my cattle. But unless they make a better job of it than they did on that occasion they'll never get me there.

For my enlightenment could you tell me please whether the case against me is being brought by the Min. of Ag. and Fish. or the Secretary of State for Wales? It is the latter whose name is given on your letter of Oct 17th. If that is the case I suppose it wouldn't be much help for me to write to my M.P.

In due course, by Order of the Court, the Secretary of State was substituted as Plaintiff and Watcyn was officially once again in contention with the old enemy, the Min. of Ag. and Fish. He was also informed that the case would now be heard even if he did not sign a Certificate of Readiness.

And at this point we come to the reference at the beginning of the chapter that you can't win 'em all.
Watcyn had been looking for the publicity for his cause which he had anticipated from a Court hearing. But, although the gentlemen of the media had had some splendid copy up to this stage, the Ministry outflanked Watcyn by having the case heard in a Small Claims Court, because the figure was under £500, and there was no press coverage.

What was worse, Watcyn was not legally represented and the various documents for which he had asked were handed to him only minutes before the case was heard. This may be referred to very loosely as British justice.

On reflection, I think he would perhaps say it was not one of his better days.

Still, fair play, they gave him time to pay.

CHAPTER EIGHT
What's it worth?

I mentioned earlier that, in 1972, Watcyn had bought a twenty acre holding at White Thorn, near Camrose. In due course he was to make certain planning applications in respect thereof. But he had found out about planning much earlier.

It is generally recognised that there are councillors, as well as officials, in this life who are corrupt. Only occasionally is it possible to prove it. Sometimes the corruption does not amount to much more than a question of you scratch my back and I'll scratch yours, and there is also the element of inborn stupidity.

I have seen it all at work at first-hand. Long ago I did a five-year stint as a member of a local authority where the Clerk subsequently went to gaol for eighteen months and the Consulting Engineer appeared in the New Year's Honours List with the O.B.E. whilst currently under notice of dismissal for his part in the fraud which was to put the Clerk behind bars.

The effective autonomy in practice of some of these Jacks-in-Office is just one of the darker facets of so-called democracy. It is revealing indeed to see their ability to manipulate a sufficiently numerous caucus of inept members who bask in their own vainglory.

When I packed it in I had occasion to write to the local

paper and said, 'All councillors are fools. Half of them have no brains and are therefore fools. The other half have brains and are therefore bigger fools for wasting time with the half who haven't. Therefore all councillors are fools.'

Maybe it was a bit of hyperbole to say that half of them had brains because, out of a total of thirty odd members, no more than five were worth their salt. But I wanted to be charitable.

The following week I was walking through town and was hailed from across the road by one of those members who was far from ranking amongst my list of five. He came over, shook me by the hand and congratulated me on my excellent letter. Then, for good measure, he said, 'When I read it I said to the missus, "Old Roscoe's quite right. I don't know how I bothers wasting my time with some of them dull buggers."' And he was one of the biggest idiots of the lot. Honest enough, but he always had to look towards one particular official for the nod as to which way he should vote.

Back in the 1960's Watcyn's father had left him a little house and a field of 1.365 acres in the village of Letterston which straddles the A40 between Haverfordwest and Fishguard.

The house was occupied by a tenant, but Watcyn thought it would be a good idea to apply for planning permission to build some houses on the field. So he went to see the Planning Officer who advised him not to apply for too many because of the question of access to the main trunk road. The fact that the road was dead straight and had recently been widened was presumably of no relevance. So Watcyn applied for permission to build one house. And it was turned down.

He appealed, of course, and was told by the Planning

Officer to put in a written appeal. When the Inspector came to look at it, Watcyn was told that he could not speak to him because he had already stated his case in writing. And the appeal went against him and his application was turned down on the grounds that the volume of traffic to and from the one dwelling would be an additional hazard on an already busy main road. How fortunate we are to have these Jacks-in-Office to protect our interests.

By this time, however, the house had become vacant, so Watcyn sold the house and field together with vacant possession. And, within twelve months, planning permission for the field had been granted. For what? No, not for a dwelling. For a garage. M.O.T. tests, second-hand cars. The lot. Maybe somebody had taken a traffic census and decided that there was not so much traffic as they had thought when Watcyn had applied for planning permission for one house, a year previously.

When Watcyn bought White Thorn in 1972 it was probably because it gave him that much extra land. A nice little river ran through it and, with his great interest in country things, he then decided to make a lake down at the lower, marshy end. Having created an attractive, one acre lake, he then stocked it with trout.

By the mid 1970's farming people were already being told about the need to diversify and what benefits there were in becoming part of the great leisure and environment industry.

So Watcyn pondered on these things and decided it would be a good idea to build twelve fishing lodges. The people who came to them would spend most of their time there, and they would be within a quarter of a mile or so of the shops in Camrose village. Therefore they would not need to use their cars to go shopping. It all looked as if it would do the local economy some good. Small wonder that the local population, as represented by the Community Council and the local representation on the District Council, supported the idea.

In evaluating these matters, however, we must always be mindful of the element of corruption or the personal antipathy of any of the Jacks-in-Office. Watcyn thought it would be sensible, in advance of the meeting of the Planning Authority, to go along to see an ancient gentleman in local government and tell him what was proposed. Watcyn had known him since his own boyhood in the St David's area and believed he should have respect for him. He was, after all was said and done, an Alderman and a County Councillor, who was drawing a fair old whack in travelling allowances even before today's bloated rates. And the old chap put his arm round Watcyn's shoulder and said to him, in Welsh, very friendly like, 'Now listen to me, Watcyn bach. Don't be like your father always was, outspoken and strong on principles and all that. You go and see Mister So-and-So and give him sixty-five pounds, and your application will go through all right.'

Watcyn did not go to see Mr So-and-So, and he did not give anybody sixty-five pounds. Nor, for that matter, did he get his planning consent. And, on reflection, all these years later, he has never quite been able to work out the significance of such an exact and unusual figure as sixty-five pounds. Personally, being of a nasty turn of mind, I reckon that the going rate was more likely to have been fifty pounds, but the Alderman would have been looking to skim fifteen off the top as his own cut in return for the introduction.

Perhaps it is all our fault for failing to understand these

characters and how their minds work.

Who said, 'I have spent the best years of my life giving people the lighter pleasures, helping them have a good time, and all I get is abuse, the existence of a hunted man.'?

It was Al Capone who said that. Maybe you've heard of him. Terribly misunderstood, poor dab.

Watcyn made that first application back in 1977. Refusal led to a public enquiry and another refusal. There was, however, a glimmer of light at the end of the tunnel, for the Inspector disposed of the fatuous objection from the Jack-in-Office from the Highway Authority about traffic congestion.

Having had this small encouragement to keep nagging away, Watcyn finally received planning consent, a mere eleven years later, in the summer of 1988. But by that time many other things had happened as we shall see later.

Initially, having had his plans for the fishing lodges turned down, Watcyn's mind had started moving in other directions and he had submitted plans for a bungalow. One day, he thought, his son would take over at Bunkers Hill and Watcyn and his wife could go to live at White Thorn. He was farming 114 acres, so there was every justification for a further agricultural dwelling. The Jack-in-Office at the Planning Authority, however, ruled that the holding was not viable. I think he kept bees so probably knew all about such matters. And Watcyn's application was turned down.

A week later consent was given for a bungalow on a ten acre holding on the opposite side of the road, to a butcher who said he wished to establish a pig farm. In the event the piggery never had more than twenty pigs. They were porkers at that, not sows. Two years later, when slurry from the piggery polluted the river and killed all the trout in Watcyn's lake, it was a good excuse to close the piggery down and the new bungalow had been built anyway, so that viability no longer mattered. There was, of course, no prosecution over the pollution but, at least, it meant there could no longer be any question of turning down Watcyn's renewed application. There were enough honest councillors to make sure of that, and no question of sixty-five pounds for Mr So-and-So entered into it.

So now he had planning consent both for his fishing lodges and his bungalow. Like corn in Egypt. Read on.

There was also one other interesting development at this time and, when reference has been made to it, it will be possible to turn from the antics of bureaucracy and officialdom to matters financial.

Once upon a time there was a rather pleasant grove of elm trees on one part of Bunkers Hill. Then Dutch elm disease struck and, having cut the dead trees down to make the place look tidy, Watcyn was left with a six acre gorge or ravine. It was not long before he had rather a good idea. If he could find something to fill in this chasm he could eventually cover it over with soil and thereby acquire some useful extra land.

It was not long before he found a most lucrative source of material. Hence the designation as an in-fill area. He found a firm who were more than pleased to dump all manner of valuable material there. They had a contract for clearing building sites, the leavings from public works contractors, oil refineries, aerodromes, offices and all manner of other places.

There for the taking, in this Aladdin's cave, Watcyn found furniture, electrical goods, beautiful tools, chains, wire ropes, stationery, industrial protective clothing, copper, aluminium, pictures, ornaments, clocks and such a variety of this and that and one thing and another as it would be impossible to imagine.

I mentioned earlier that, as a youth, Watcyn had learned his trade as a butcher. For some time at Bunkers Hill he had been turning this training to good use and had built up a useful little business delivering meat produced on his own place. Now, the butchery had to take second place as he endeavoured to pick out the best from each load at the in-fill area before the next load arrived to be dumped on top of what was already there. And the revenue from this source was all entered in the farm accounts.

The entire business was above board, but this did not prevent the Jack-in-Office from Environmental Health from taking a sudden interest. All along the line he had not wanted to know about Watcyn's legitimate and pressing pollution problems, but now, when he could see Watcyn set fair to make a few pounds, it was a different matter. He tried to say that Watcyn had a refuse tip and needed authority to operate it. But Watcyn did not have a refuse tip, or anything like it. He had an in-fill area. Then, one day, the Jack-in-Office paid a visit and Watcyn said very well they could go across the fields on the tractor. And just to show how stupid this particular Jack must have been he got up in the tractor cab with Watcyn. And you know all about this particular tractor cab because you have read a whole chapter on the subject. And Watcyn drove Jack up a steep bank and then down a steep bank and put the fear of God into him on the grand scale. After that Jack seemed to grow suddenly much less enthusiastic.

Then, just before Christmas 1986, Watcyn found something on the in-fill area which was rather out of the more usual run-of-the-mill deposit. He had been out rabbiting by night with the dog and a lamp and, as he was on his way home by the in-fill area, a piece of paper blew up against his chest. He shone the lamp on it and was more than passing interested. As a result, he went back in daylight the next morning and found about a dozen plastic bags of documents which had been dumped on behalf of the Haverfordwest branch of the National Westminster bank.

And that, as the saying goes, is where we came in.

CHAPTER NINE
Rising cost and papers found

When Watcyn took on the tenancy of Bunkers Hill in February, 1969, he was banking at the Haverfordwest branch of Barclays bank and had an overdraft of £1,500. A year later he had an agreed limit of £4,000 and the records show that he owed the bank £3,358. He would seem, therefore, to have been well within his limit.

These figures are quoted from Watcyn's own statements, most of his records at the bank having disappeared.

Shortly before this, as we have seen, Watcyn had inherited, and then, following the planning nonsense, sold, the little property at Letterston. He sold it for £8,000 or thereabouts. In conformity with Mr Denis Healey's expressed intention of squeezing the private sector until the pips squeaked, £3,000 had to be paid in Capital Gains Tax. The remainder Watcyn invested in a building society. The handling of this business was left to his solicitor.

Some years later Watcyn entered into an agreement concerning the purchase of a hedge-trimmer. Then a gentleman came along to query the deal since Watcyn seemed to be one of these disreputable types with his name recorded in some Gazette or other because he had been in court for his failure to pay a sum of £3,000. The figure was in respect of the Capital Gains Tax and it was the first intimation Watcyn had that he should have been anywhere near a court, the gentleman from the hedge-trimmer outfit expressed himself as being quite happy, and Watcyn was told by a third party that for a payment of fifty pence his name could be removed from that list. So Watcyn telephoned his solicitor who said that was all right, boy, and leave it all to him and it still hasn't been done.

In 1972 Watcyn bought White Thorn for £6,050 and, with the bank's approval, paid for it out of his current account.

In 1977 he entered into discussion with his then landlord as a result of which he agreed, as a sitting tenant, to purchase Bunkers Hill for £24,000. At that stage his overdraft stood at £9,000, which figure included the £6,050 paid for White Thorn five years previously. From which it would seem that, up to that stage, he had been acting quite prudently in his business affairs.

Having entered into negotiations to buy Bunkers Hill he went to see the bank manager with a view to arranging the loan through the Agricultural Mortgage Corporation on a fixed interest and repayment basis. The manager told him to do the deal with his landlord and then come back and discuss the financial arrangements. By that time, he said, a new manager would have taken over.

Once he had clinched the deal Watcyn went back to the bank where the new manager persuaded him that Barclays could offer him better terms than the A.M.C. and they advanced the money. There seems to have been an agreement that the position would be reviewed in twelve months time with a view to going to the A.M.C. if their terms were more favourable, but this will be referred to later.

Watcyn provided £5,000 of the purchase price himself by withdrawing the money he had invested in the building society, and the bank also took the deeds of White Thorn as additional security. The bank were, therefore, providing £19,000. Watcyn's monthly payment amounted to £256.50. This included interest and capital repayment with the understanding that the capital would be repaid by the end of ten years.

Having acquired the freehold of his farm Watcyn then proceeded with a number of improvement schemes, all of which were grant-aided by the Ministry of Agriculture. On every occasion he did nothing without consulting the manager and obtaining the bank's approval. And most of the schemes were carried out on a standard cost basis, so that Watcyn was doing much of the work himself and by direct labour, which meant that his indebtedness to the bank should not have been increasing too seriously.

In the spring of 1984, however, he saw from the annual balance sheet prepared by his accountant that, for the first time since he had started farming, he had operated at a loss on his year's trading. So he sat down with his wife to see what they could make of it all and what economies could be effected, and they were horrified to realise that the loan account on the purchase of the farm was still nearly as high, seven years later, as when it had started, even though the monthly payments of £256.50 had gone through regularly. Additionally, the overdraft had rocketed from £9,000 to £111,000.

By this time there had been another change of manager. He could not give Watcyn any explanation as to why he still owed nearly as much on the original loan as when he had started but said there must be an agreement somewhere. For the next eighteen months, at least once a month, Watcyn went into the bank in Haverfordwest to try to sort things out, and every one of these visits was at Watcyn's instigation and not by the invitation of the manager. Eventually the manager said they could not find any agreement and then admitted, or claimed, that all Watcyn's papers had been lost.

During one of these many visits to the bank the manager asked Watcyn if he would like to have a talk with the loans officer who was there that day and, of course, Watcyn was only too willing to talk to anybody who might be able to shed light in dark places. So the loans officer came in and Watcyn asked him how long it would take him to pay off the loan on the farm.

The loans officer looked at the sheet of figures produced by the manager, did a few calculations and said, 'You'll never pay it off.' So Watcyn said, 'Would you say ten years, fifteen years, twenty years? How long? Name a figure.'

'You'll never pay it off,' the man reiterated, and Watcyn said, 'Thank you very much, gentlemen,' and walked out.

Some years previously Watcyn had paid a firm of civil engineers a fee of £1,000 to carry out a survey on part of his farm and they had confirmed that there were valuable deposits of sand and gravel there. Reputable agents valued the twelve acres involved at £120,000. In 1984 Watcyn made this twelve acres available for sale and certain public works contractors showed considerable interest.

In September of that year, however, six months after Watcyn had begun to try to sort things out with the bank, he had a telephone call one lunch time from a friend of his in the haulage business who said he was speaking from the Cleddau Bridge Hotel, near Pembroke Dock, where some sort of meeting or conference of public contractors and haulage firms was being held. And this friend told Watcyn that he could forget about any hope of selling his sand and gravel for a fair price because his bank manager was going round offering it to various people. And the word was quickly whispered abroad that Watcyn Richards was in financial trouble and the vultures could prepare to move in to pick the carcass. Sure enough, two days later, Watcyn had a telephone call with a derisory offer from a firm who had previously expressed interest at a much higher figure. It did nothing to improve Watcyn's relationship with the bank manager.

By that time, with more and more interest accruing, the overdraft stood at £128,000, and the manager had started to intimate that Watcyn's papers had been lost.

The matter then dragged on for another twelve months until eventually, in September 1985, after much pressure from Watcyn, he was able to have an interview with Barclays' local director in Cardiff. And this is where we come to what I referred to earlier as the much more sinister business which was the reason for Watcyn cancelling his membership of The Onion. Two office holders accompanied him on his visit, and Barclays knew that they would be coming because they had insisted on being told by Watcyn beforehand who, if anybody, would be accompanying him.

And he had intimated that he would be represented by these two characters. Watcyn drove them up in his car but, as they were nearing Cardiff, one of them broached the subject rather tentatively and said that they would only be there as observers and would not be representing him. So, when he could get no more sense or satisfaction than that out of them, he pulled into a lay-by, stopped the car and told them that, if they couldn't tell him what was behind it, they could get out of the car and then so-and-so well walk home. And rather than be dumped by the roadside they told him.

A telephone call, they said, had been received from Barclays the previous day to say that, if The Onion supported Watcyn, then Barclays would not support The Onion in a case in North Wales where they were themselves currently involved in a large claim and were relying heavily on Barclays for support.

So that explains about The Onion. We make few friends in life by saying, 'I told you so.' When Watcyn reported back I contented myself by saying, 'Blessed is he that expecteth nothing, for he shall not be disappointed.'

To all intents and purposes, therefore, Watcyn was on his own. As he went in to the local director's office he was confronted by a character sitting at the far end of 'a big table the size of a football field,' and, when he turned round, saw that his bank manager was standing behind the door.

The only thing he could say afterwards in all fairness was that the manager had told him it wouldn't do him any good going to Cardiff. The local director said that they would now amalgamate the two accounts. Although Watcyn did not agree to this or accept it, nevertheless, entirely against his expressed wishes, it was done. By this time, of course, interest had been piled on interest and the total figure now stood at £171,000. I also have a note in the bundle of papers before me as I write of a figure of £156,005, but I cannot think where that came from, and the whole business had by this time reached such heights of unreality, and was so far beyond the comprehension of ordinary mortals, that I hardly think it matters. Otherwise I would have to start worrying and puzzling about that odd fiver and where and how that came into it.

There was one other outcome of this Cardiff visit. The bank arranged for their agricultural adviser to visit Bunkers

Hill so that he could put Watcyn on the right lines. It was raining on the day the adviser called and he refused to go down to the field to see the cows because he did not want to get wet. Even so, he felt competent to prepare a report.

I hope Watcyn makes sure that it is preserved for posterity because it was the most fatuous load of rubbish imaginable.

Later on Barclays were to argue that they do not reckon to give advice. Based on this joke document they could never have said a truer word.

During the eighteen months that this had been going on, apart from all the other business of antibiotics and Ministry of Aggravation and vexations various, there had been the challenge to the farming industry of the introduction of milk quotas. One of the Government's leading 'Wets', Peter Walker, as Minister of Agriculture, had made a pronouncement to the effect that farmers should go on producing milk for all they were worth and that the country needed every pint. Prior to this, as Ted Heath's henchman, he had been largely responsible for foisting Dyfed and many other such abominations upon the people. Having said his piece about milk, he was then himself foisted by Mrs Thatcher upon the people of Wales as Welsh Secretary to show the high regard in which she held us. Hard upon his departure from the Min. of Ag. and Fish. and Food dairy farmers were told that they must cut down on milk production and, overnight as it were, quotas were introduced.

This tendentious measure was inflicted upon the highly efficient dairy producers of this country as part of the E.C. set-up to assist the producers of lower quality milk on the Continent. When quotas were introduced there were many cases of hardship, particularly among the younger farmers who were building up their herds and those who had been rearing heifers which were just due to come into the herd. Watcyn came into this category, having been building up his herd again following the earlier disasters already fully chronicled in these pages.

Tribunals were set up to consider such cases of hardship, and Watcyn was one of those who had to appear.

As a neighbour of mine at the time said to me afterwards it was a bit like having to go to the dentist, except that at the dentist they gave you anaesthetic. He was also kept waiting for two hours before being ushered into the presence of the tribunal, whose members were supposed to know all about efficiency, because they were already running late in the conduct of their business. And if you were being paid ninety pounds a day plus expenses maybe you would contrive to run two hours late as well. Then, said my neighbour, when he was eventually ushered into their presence he found himself confronted by a retired bank manager, a solicitor and an agricultural adviser. 'And these,' he said, 'were the buggers who got us into the mess in the first place. And what would they know about hardship anyway?'

Watcyn fared no better than hundreds of others.

To return now, then, to matters financial in general and Barclays bank in particular.

Having returned from his fruitless visit to Cardiff, Watcyn went to see his solicitor. The best advice he could offer was to sell up, but he also promised to get counsel's opinion and to see about getting legal aid. As it transpired he did not fulfil either of these promises.

Early in 1986 Watcyn had a letter from the bank to

say that a cheque he had paid out was being returned. The cheque was in respect of his month's feeding stuff, so he immediately telephoned the firm concerned to say he would look into it immediately, only to be told that the cheque had been honoured and what was he talking about. A day or two later the same thing happened with another firm. Notification came from the bank that the cheque had been returned although it transpired that it had been paid. Watcyn went along to the bank, but the manager was not there and he saw the second man, as I believe these characters are called. He said to Watcyn that he must know what his agreed limit was and obviously he must have exceeded it. Watcyn said he didn't know anything about any agreed limit, whereupon the second man went for Watcyn's file, looked at it and expressed himself as being completely bewildered.

About this time the manager retired. He was a good sort and I liked him very much. In spite of all that was happening Watcyn liked him, and he often confided to Watcyn that the job was getting him down. Not long after he retired he died suddenly and there can hardly be much doubt that the stress and strain contributed to the cause.

When Watcyn saw the new man he took an instant dislike to him and I know several people of the same opinion. The nonsense of the cheques which were allegedly being returned but were, in fact, being honoured, continued. And then Watcyn noticed from his bank statements that each time it happened a service charge of fifteen pounds was being made. No wonder country branches of banks throughout the land have been transformed into glass-plated, mahogany-lined and deep-upholstered, computerised palaces.

There was only one thing for Watcyn to do, and he did

it. He opened an account, in credit, with the Haverfordwest branch of the Trustee Savings Bank and issued instructions to the Milk Marketing Board please to pay his monthly milk cheque direct to him. Then the what-not hit the fan. That is to say, in other words, Barclays didn't like it.

If you look at it from Watcyn's point-of-view you will realise it was the only move he could make. As long as the milk cheque continued to be paid into Barclays he would be entirely at the mercy of any bloody-minded action the new man might care to take. And it would only be prolonging the agony as the whole miserable business headed towards its inevitable end with foreclosure as the only outcome.

Vowing a vow to have no more truck with bank managers, Watcyn was content for his wife to go in with the first deposit and collect the forms for signature from the T.S.B. along with a paying-in book and cheque book. The new manager at Barclays got on to the T.S.B., and the manager there sent word to Watcyn that he wished to see him immediately, pronto and forthwith. To which Watcyn replied that he had no intention of going anywhere near a bank manager again for the rest of his life.

So the T.S.B. manager said he was sorry but he could not allow such a situation to continue, and Watcyn said that was fair enough and that he would be more than happy and willing to take his money somewhere else. And that was the end of that and no more was said about the matter.

He paid all his bills as they came in each month and, as the surplus began to grow, he began to realise how he had been contributing to the swelling of Barclays' coffers over the years.

That summer there was what Mr Nigel Lawson would no doubt have referred to as a slight blip in the smooth running of financial matters when the Customs and Excise people paid Watcyn's V.A.T. into Barclays in spite of his having asked them please to pay it in to his new account at T.S.B. They wrote and apologised on July 3rd, saying, *'The repayment of £626.24 was sent to your bank in error. This will be recovered and forwarded to you at the earliest opportunity by payable order.'*

A month later they wrote again to say that they had been on the telephone to Barclays to discuss the matter and it had been agreed that Barclays would not be returning the money but would be crediting the amount to Watcyn's account.

So Watcyn wrote to the head wallah concerned in the business, thanked him for his letter and said, *'I am not interested in any 'phone calls you or yours may have made to Barclays or anything those rogues may have said.'*

He then recapitulated on the question of the undertaking given in the previous letter with regard to recovery and repayment and said please would they now expedite this undertaking forthwith, or words to that effect. And, fair enough, they expedited forthwith, which is another way of saying they arranged for onward transmission. In fact, they paid him.

Whether they ever recovered the money from Barclays, or wrote off the payment to them as a loss, we shall probably never know, but one interesting observation may be permissible.

In the fulness of time Barclays were directed by a Court Order to produce the records which all along they had been telling Watcyn they had lost. Wherefore and therefore they had to unlose them and an entry for August 12th 1986 is of interest.

Only movement on account since excess report on

30:6:1986 has been a V.A.T. payment, probably by mistake which we have refused to refund in spite of requests.

The inside story could possibly be of interest, but all that concerned Watcyn at that time was that he had been paid.

By the autumn of 1986, Barclays' new manager was threatening to foreclose and Watcyn was still asking to see the records which were supposed to have been lost. A few weeks before Christmas, 1986, Watcyn received notice to quit. Shortly after that he ran into the business of the documents which had been dumped for the Nat West. Anyone who has read thus far will have a much better understanding of Watcyn's state of mind than ever they could have had merely by reading the reports which were shortly to appear in the papers.

When he had the notice to quit he went to see his solicitor and, with his mind in turmoil, he agreed to sell White Thorn, his cows and his milk quota, which would just about clear his alleged debt to Barclays and leave him with Bunkers Hill unencumbered. No action was taken immediately and the advertisement had not appeared when Watcyn chanced upon the documents dumped on his land on behalf of Nat West.

Although there is still much to be written on the subject of Watcyn's saga with Barclays, much of which will explain his state of mind at the time, since it immediately preceded the finding of the Nat West documents, it would be as well to dispose of the Nat West interlude at this point since they were very much an innocent party to something not exactly of their own making, however much they may have been at fault over the slip-shod handling of highly confidential documents. It was their misfortune that, because of Watcyn's dealings with Barclays, he held banks generally in such odium. Barclays were still claiming they had lost his papers and, for all he knew, they could have done the same with them as Nat West had done with those relating to some of their clients. Many of the letters and accounts relating to some of their clients whom the world believed to be financially secure made interesting reading. Solicitors, had they known that details of their nine million pounds of clients' money were being blown about the countryside, would not have been amused. Some of the internal memos and correspondence on this subject were, to the outsider, amusing, highlighting as it did the truth of the well-worn saw that great fleas have little fleas upon their backs to bite 'em, and little fleas have lesser fleas, and so ad infinitum.

We are all answerable to somebody somewhere along the line. Even bank managers. We all have somebody on our backs.

As we have seen, having, just before Christmas, been served with notice to quit and agreed with his solicitor to sell White Thorn, his cows and his milk quota, Watcyn had plenty to occupy his mind over the festive season whilst solicitors and those who earned their daily crust in banks were taking protracted rest from their labours. In the New Year Watcyn went to see a well-known business man he felt he could trust. If I do not name names, may I explain that there are already no fewer than three characters in all this by the name of David Jones and, for good measure, the Nat West manager was another Jones. So that makes four of them, even though they may not have been as well known as the one in the classic Welsh story of the Pope and Mr Jones of Aberystwyth. But rather confusing, even so.

Suffice it to say that the business man, who banked with Nat West, told Watcyn he would go to see the bank manager and telephone Watcyn that evening. He did not telephone Watcyn that evening but, the following morning, which was January 6th, 1987, he turned up at Watcyn's farm in the company of the bank manager and his assistant.

When the manager saw some of the documents his face, not surprisingly, turned white. Making due allowance for the fact that he must have been in the same state of near-vomiting as many of the banks' customers so often are when the shoe is on the other foot, the fact remains that in no time at all he was assuring Watcyn that he knew all about his troubles with Barclays and that his bank would sort all that out and take over the account and don't lose any more sleep over it but to come in and see him. This, as I said earlier, was in the presence of a highly credible witness. And the same highly credible witness went into the bank with Watcyn a day or two later only to find that the manager had changed his tune and was now telling Watcyn that the Nat West would pay his out-of-pocket expenses.

On this occasion Watcyn was unable to give a detailed statement of his affairs, but promised to set it out in writing, which he did shortly afterwards. In the same letter Watcyn said that, as a gesture of good-will, he was willing to allow the manager to come to his in-fill area to collect the considerable quantity of documents which were still there and which Watcyn had not bothered to collect.

Although he was quick enough to take Watcyn up on this offer, and turned up with an assistant and another character driving a land-rover, Watcyn did not like his attitude.

Amongst other things, whilst out of earshot of his two companions, he told Watcyn that where they had troublesome characters to deal with they were quite ready and willing to 'send in the heavy gang.' At the time there had been a story in the national press of the Nat West having brought in a 'hit-man' from South Africa but, when he started threatening Watcyn, the manager had really made a serious error of judgement.

That same day Watcyn wrote a further letter to the manager telling him that he was surprised at his attitude, mentioning the possibility of a reward rather than out-of-pocket expenses, and further suggesting that it would be as well to report the matter to his head office. Watcyn also, at the manager's request, arranged for a J.C.B. to cover in all the loose material because there were still so many documents mixed up with it, and the arrangement was that Nat West would foot the bill. But the manager reneged on that one as well, and Watcyn was lumbered with the bill.

The explanation being offered by Nat West for the presence of sacks of their documents on Watcyn's in-fill area was that the bags had been picked up in error, along with two old chairs, by builders who had been working at the bank. The fact remained, however, that they had not been properly taken care of by the bank and had finished up in what was to all intents and purposes a public place. And whichever way they had found their way there would have been of little interest or consolation to those whose personal and private affairs could so easily have become common knowledge.

Watcyn then had a letter from the Nat West's solicitors in London demanding the return of the documents in his possession and saying that their clients were more than willing to arrange collection.

Although Watcyn had agreed with his solicitor to realise on some of his assets he had, understandably, been loath to take the final steps, but his solicitor now pressed him to go ahead with his plan, and this he did.

Before doing so, however, he replied to Nat West's solicitors, pointing out certain facts of which they seemed to be unaware, agreed that they could arrange for the collection of the documents, and asked them whether they would like to have the broken chairs back as well.

Then everything seemed to happen at the same time. Having been misled in the first place, Watcyn was still fostering some foolish notion of getting something out of it. The term bank manager, to him, was all-embracing and, as he was in the position of being slowly crucified by Barclays, I, for one, find it hard to blame him. But not being sufficiently experienced in these matters, he did not understand about publicity. It is a sharp weapon indeed, but it blunts very quickly. It can only be used once. The fear of publicity is a much more potent weapon than the publicity itself.

Seeing which way things were going, and with Watcyn's knowledge, I had spoken to a very experienced journalist friend of mine. He had recognised that the story was dynamite and had agreed to bide his time. We were both concerned only with the situation developing in a way which might possibly help Watcyn. That was on Jan 24th.

On Jan 27th Watcyn saw his solicitor and agreed to the advertising straight away of White Thorn, his cows and his milk quota. As we have seen, the thinking was that the sale of these assets would clear his alleged indebtedness to Barclays and leave him with Bunkers Hill unencumbered, including the valuable sand and gravel deposits, as well as his remunerative in-fill area. It was that same evening that he wrote to Nat West's solicitors telling them that they could collect the documents which were still in his possession.

On Thursday, Jan 29th, Watcyn saw the advertisement in the Western Telegraph and it gave him a sick feeling. There it was in cold, harsh print. Everything for which he had saved and struggled going up in a puff of smoke, as the bitterness of the awful truth came home to him.

Unfortunately, I was out when he telephoned me and then, with no restraining influence, he contacted my journalist friend who, at Watcyn's behest, gave the story to the Sunday Express. It was an utterly stupid thing for Watcyn to have done and I told him so in no uncertain terms, and also told him he could now whistle goodbye to any hope of financial reward. He would have his publicity and much good it would do him.

On the Friday the Nat West bank manager turned up bright and breezy to collect the documents, but Watcyn said, 'I've changed my mind.'

'You can't do that,' said the manager. And Watcyn said, 'I've just done it.'

Later the same day the Sunday Express photographer turned up but Watcyn refused to cooperate. He wanted any picture being taken to show the documents being handed over. He had an idea that this would draw attention to the alleged losing of his own papers and records by Barclays, for it had not been established at that point that Barclays were lying in their teeth, as an old schoolmaster of mine used to say.

That evening, Friday Jan 30th, Watcyn wrote a further letter to Nat West's solicitors:

When I wrote to you on Jan 27th and told you that your clients could collect the various documents I assumed they would have the courtesy to make an appointment to come at my convenience like customers have to do before they may be ushered into the presence of their bank manager. In any case, as I have already told Mr J. O. Jones, these documents have been buried in various safe places and it will take hours to gather them all together.

I was infuriated this morning, therefore, when Mr J. O. Jones and a couple of his minions turned up on my farmyard unheralded and looking like the cat that swallowed the canary. So I said I had changed my mind and I believe I have as much right to do that as your Mr J. O. Jones over his original intimation to see that I was financially recompensed.

What I have decided now is that you must make an appointment for them to come to collect the various documents, and I shall ensure that the handing over be done in the presence of whatever press and camera men may be interested.

I shall await your reply.

The solicitors would not have received that letter until the Monday, or possibly the Tuesday, but any good it may have done had been completely and utterly nullified by the appearance on the Sunday in the Sunday Express of the story which I quoted in full at the beginning.

Immediately Watcyn's telephone started ringing as he received calls from all over the country from people with grievances, real and imaginary, against their banks. One man from Grimsby, or was it Hull – I cannot just recall, except that it was one of those fishing ports – telephoned Watcyn and told him to return the documents to the various people concerned. Had he known then what he was to discover later he might just as well have done it.

On the Monday Watcyn's place was swarming with reporters and photographers and it was at breakfast time that day, as I recall, I jokingly said to my wife that I ought to write a book about Watcyn. There is an old saying that even a fish wouldn't get caught if he'd only keep his mouth shut. I realised, too late, that I should have kept my mouth shut, because there is many a true word spoken in jest. My wife thought it was a splendid idea, so maybe she didn't know about Huckleberry Finn.

About midnight on Tuesday, Feb 3rd, Watcyn was served with a High Court writ. The wording was a complete cock-up and the man had to come back again a few days later, but that is incidental to the story.

I mentioned, way back in chapter five, that a friend of mine had said when in his cups that he wouldn't lift a finger to help Watcyn so-and-so Richards. But he was himself fighting a moderately successful rearguard action against Barclays at the time and he was man enough to say, 'You tell Watcyn Richards not to sign anything for Barclays.'

The day after the writ was served on behalf of Nat West Watcyn took it to his solicitor who said to sign certain documents for Barclays regarding the sale of White Thorn, his cows and his milk quota. But Watcyn, who had needed little, if any, prompting said he was not signing anything. So the solicitor said in that case he could no longer act for him, and there was Watcyn stuck with the writ from Nat West and with no legal representation. To find a new solicitor was suddenly top priority.

As it became increasingly apparent that he had not had the best of legal advice so far, a young solicitor took him on and, up to a certain point, did very well indeed for him. The immediate priority was to obtain legal aid to handle the High Court injunction.

A couple of years previously, following his abortive visit to Barclays' local head office in Cardiff, Watcyn had been to see his solicitor who expressed himself as being dissatisfied with the set of accounts prepared by Watcyn's accountant and asked for a fresh set in a different form. When these also failed to meet his requirements he told Watcyn that he would find him another accountant. Now, when Watcyn had to find a new solicitor, this new accountant also said he could no longer act for him. It was the only useful or positive thing he ever did for Watcyn. But, first of all, the business of Nat West had to be disposed of and the matter was urgent.

In his disturbed state Watcyn was not, in my opinion, helped overmuch by letters, telephone calls and visits from well-meaning people, many of whom had suffered grievously at the hands of the banks. Whether he would want to take on the world in due course was irrelevant. All that mattered for the moment was to extricate himself from the hole he had so foolishly dug for himself by going to the Sunday Express in such an untimely manner. There was a possibility of damages and costs against him and, if Barclays were not going to bankrupt him one way, the Nat West could most certainly do it another. All of which was utterly stupid when Nat West were in no way responsible for his predicament and there was never any reason why he should have become involved with them.

One of the more active of the people to contact Watcyn when the story appeared was a man by the name of John Pett, who had had some miserable experiences at the hands of his bank, and who had now formed an organisation which he called The National Association for Victims of Fraud and Banking Malpractice. As the old saying goes, 'It's money makes the donkey gallop.' But the people interested in John Pett's organisation were those who, in one way or another, had been robbed by the banks, as a result of which they had little money left to make the donkey gallop very far or very fast. So, what with one thing and another, I reckon they have an uphill struggle on their hands.

The writ was returnable, if that is the correct term, on Tuesday, Feb 10th, and there was to be no nonsense about college lecturers telling us how to get there or where to park. We went by train to stay overnight in London and were met by John Pett that evening. He is a man who has suffered, but his story is not for telling here.

In the morning we went by taxi to Lincolns Inn, having calculated that for three of us it would be more sensible and expeditious than either the underground or bus. Watcyn paid the taxi driver and, as he caught up with us, I asked him how much the fare was. It was five pounds. So I said, 'How much tip did you give him?'

'I didn't give him a tip,' he said.

Maybe I looked a bit surprised, so he added, 'Nobody ever gives me a tip.'

'Quite right,' I said. 'Quite right.'

Not exactly the sort of man to give sixty-five pounds as a dab in the hand to some corrupt character from the planning authority.

We met the young solicitor deputed to act for Watcyn in London and then we met the young lady barrister deputed by the solicitor. There were various characters sculling around dressed up in wigs and gowns and all this and that. Then John Pett declared his hand and the Nat West's legal eagles immediately said they were objecting to anybody else going in. Whether they were within their rights or not I did not know and cared even less. My only concern was to see it over and done with.

The ruling was that the documents should be returned to Nat West for safe-keeping until such time as ownership could be decided. I reckon that maybe that was an odd sort of ruling considering how the documents came into Watcyn's hands in the first place and that, if British justice were all it is cracked up to be, and they were really concerned about safe-keeping, the Court should have taken charge of them. Especially as there are far more builders' skips in London than ever there are in Haverfordwest. Come to think of it, poor Mr Norman Lamont can reckon he was lucky that he did nothing more reprehensible than buy a couple of bottles of plonk at a London off-licence.

That was, to all intents and purposes, the end of the case of Nat West versus W. Richards, and it cleared the decks for the real fight which was, as it had been from the start, with Barclays.

CHAPTER TEN
Not a satisfactory settlement

It is impossible, rather than merely difficult, to place the various events at this stage in chronological order, because so much was happening at the same time.

The more discerning reader may already have noticed certain discrepancies in the time scale because the writing has taken place over a protracted period. Apart from a sojourn in hospital followed by a time when writing was consequently out of the question, there is even yet, at the time of writing, a fair old volume of water up-river which has yet to come down-stream before it flows under the bridge.

It was towards the end of February 1987 that I began writing. As I write this it is June 1989, and impossible at this stage to know what the eventual outcome will be, so that all I can do is take one step at a time. Like Cardinal Newman, *'I do not ask to see the distant scene, one step enough for me.'*

In due course Barclays were ordered to produce the missing papers, but it is still open to question as to whether they produced all of them. Even so, they produced enough for writing and comment on them to fill a book in itself. One thing which is abundantly clear is that the reports prepared by the Haverfordwest branch were strictly for the benefit of

Local Head Office and, in some instances, bore no relationship whatsoever to the true position or to what was happening. On a number of occasions the reports suggest that the bank was acceding to Watcyn's request for further financial accommodation when, the truth of the matter was, they were pushing money at him and encouraging him to carry out more and more improvements. Some of his visits to the bank were not even recorded.

One particular example of complete misrepresentation and concealment of the facts occurred on July 17th 1984. By that time the bank had been telling Watcyn that his papers had been lost, and his wife found this hard to believe. So, on July 17th, Watcyn visited the bank in the company of his wife. Under pressure from Watcyn, the manager again repeated the assertion that his papers had been lost and Ray, Watcyn's wife, could hardly believe the evidence of her own ears. The report to Local Head Office on this visit made no mention of any lost papers or of Watcyn's anger on this occasion. The report said, *'Mr Richards interviewed together with his wife following our letter of concern regarding the excess position. A full report on the Sand and Gravel project at Bunkers Hill has now been received from Thyssens and Mr Richards will now place the farm on the market by asking various Sand and Gravel companies to tender.'*

Any hope of a sale of the sand and gravel was, as we have seen, well and truly fouled up by Barclays' manager. And it is worth mentioning that at no time had Watcyn put down the sand and gravel as an asset in his statements to the bank. It was the manager who was constantly getting excited about it.

The final reference to missing papers and information would seem to be for October 27th 1977. In the epitome of arrangements and correspondence, with which Watcyn was eventually supplied, the entry for that date reads, *'£8,580 Dr. Form 21 for overdraft limit of £14,000 and TL of £19,000 for 12 months.'* This was in respect of the purchase of Bunkers Hill, but the remainder of the line and the next two lines have either been deliberately obliterated or are otherwise completely indecipherable.

Watcyn maintains that, at one of his many subsequent irate interviews at the bank, the manager eventually and very begrudgingly read out an entry from the bank's ledger relating to this transaction which stipulated that Watcyn would be called in for an interview in twelve months time to decide whether he would be better off with a mortgage from A.M.C. He was not written to on the subject again and was never called in for such an interview. And the bank have never been able to produce a copy of any letter purporting to have been written concerning this point.

The bank insisted that such a letter would have been sent in September 1978. On October 5th 1986 Watcyn wrote and, amongst other points, asked for a copy of the letter. That request was ignored and, rather conveniently it would seem, this salient point was never mentioned in subsequent correspondence.

Watcyn should have been called in to discuss the position. And, in view of rising interest rates at that time, his monthly payment to the bank should have been increased if the loan were to be paid off in ten years as originally intended.

For it to be seen what the calibre was of some of the professional advice and service which Watcyn was receiving, it might be mentioned here that it subsequently transpired that his new accountants, who had been nominated by his original solicitor, had not presented accounts to the

Inland Revenue for two years, although they had been in possession of all his papers and had charged, and been paid, for doing the work. Be it said, too, especially when there has been so much criticism of bureaucrats and parasites in general, that the oft-maligned Inland Revenue behaved very reasonably when the position became clear. These accountants were also, be it remembered, the ones who had said that they could no longer act for Watcyn when he had been obliged to find a new solicitor.

One of the first moves by the new solicitor was to suggest to Watcyn that he should ask new accountants to prepare a statement which would enable him to obtain a loan elsewhere and get Barclays off his back. Watcyn found new accountants all right and, having examined his accounts, they said that his business could service a loan of no more than £35,000, which, however sensible, was not immediately helpful.

When the case came up for hearing at the County Court at Haverfordwest in March 1987 it was obvious that, on all the evidence available at that stage, Watcyn had no hope whatsoever. A barrister had been briefed, an offer was made out-of-court and was accepted by Barclays solicitors. It was that Watcyn should sell his milk quota straightaway and pay the money to Barclays, and also put the sand and gravel on the market. The case would then be heard in eight weeks' time.

There was no reference to the cows and it should be mentioned for the benefit of the non-farming reader that, without a milk quota, the cows would not be of much use, in spite of Peter Walker's valedictory announcement as Minister of Ag. and Fish. and Food that the nation wanted every pint of milk the farmers could produce. It did mean, however, that Watcyn was free to sell the cows and reinvest the money in beef cattle and sheep.

Watcyn sold most of his milk quota for £55,000 and the money was paid to Barclays.

Before matters could proceed as far as this Watcyn's solicitor and barrister had to be paid, and it was before it had been confirmed that Legal Aid would be available. Fortunately, he had the money for his milk which for some time had been diverted from Barclays grasping voracity. He was in the throes of trying to fill in the application form for Legal Aid when he had a 'fare-thee-well' letter from his original solicitor enclosing his 'very reasonable' account for the modest sum of £529, which included V.A.T., and I always think that the reference to V.A.T. is the final irony.

This was another of those cases where Watcyn could have been forgiven for thinking that the value aspect had been rather nebulous. In his own 'fare-thee-well' letter in reply he intimated that he did not feel disposed to acknowledge the debt:

The more I think of all that has happened and is happening now, the more I feel that you have given me some poor advice and have been negligent in your handling of my affairs.

For example, I am now struggling to fill in the forms applying for Legal Aid and I feel that was something you should have tackled for me when the trouble with Barclays started. Nor did you obtain counsel's opinion which you said you would do.

More recently you wrote to me on Feb 6th. That letter was posted by second-class post and did not therefore reach me until the Monday morning of Feb 9th. That was not going to be of much help to me, unless I had made other arrangements, when I was due to appear in the High Court the following day. (This was in respect of the Nat West case.)

Following the legal activity and the ruling in February, the adjourned hearing was fixed for July 1st. Then it transpired that Barclays had changed their solicitors. Although it was perhaps because they felt that the London boyos would have a better grasp of such matters than the local firm they had been using, it was interesting to note that to shuffle the cards in this manner was not necessarily Watcyn's sole prerogative.

Legal Aid having been obtained, Watcyn's barrister offered the thought that it would be helpful to Watcyn's case if he could find a bank manager who would be prepared to examine the papers with a view to expressing the opinion that Barclays were in any way at fault. Such a manager, recently retired, was found without any trouble. Had he ever been called to give evidence he would have been seen to be a sound man and a first-class witness. He was from good farming stock and had spent the forty years of his banking life in 'Farming Branches.'

In his statement he outlined his credentials and then itemised the various dates and lists of borrowings as already set out in these pages. The remainder of his statement is, I believe, worth quoting in full.

4. I know that the Agricultural Mortgage Corporation offer both fixed rate loans and loans at a fluctuating rate of interest. I do not think that in 1977 Barclays Bank offered fixed rate loans. If I had been dealing with Mr Richards I would have made sure that Mr Richards understood the difference between a fixed rate loan and a loan at a fluctuating rate of interest. I consider that the failure of the then Manager of Barclays Bank, Haverfordwest, to do so was a breach of his duty to Mr Richards.

5. I would not have advised Mr Richards to take one loan in preference to another. I would have advised Mr

Richards to consult his Accountant because his Accountant would have had a more detailed knowledge of Mr Richards' tax position. I would also have advised Mr Richards to consider with his Accountant the advisability of taking out an endowment policy or a mortgage protection policy or both with a reputable insurance company.

6. If Mr Richards had told me that he did not wish to speak with his Accountant then I would not have insisted that he should do so having satisfied myself that I had explained to Mr Richards the difference between a loan at a fixed rate of interest and one at a variable rate of interest.

7. If I had been Mr Richards' Bank Manager then at the same time I agreed to lend the money to him and his wife to enable them to purchase Bunkers Hill I would have agreed with Mr and Mrs Richards some provision as to the repayment of the debt. I would probably have allowed Mr and Mrs Richards two years grace without any re-payment of capital to enable them to increase their livestock and begin to obtain a return on the capital invested.

8. The loans to Mr and Mrs Richards should have been reviewed, at least, annually. I would have drawn their attention to the effect on their borrowing of any increase in the rate of interest at the interview which would have taken place shortly after the annual review.

I note from the bank statements which have been shown to me that the interest payable half yearly in respect of the loan had gone up from £946.62 in June 1978 to £1,443.56 in December 1979. This had the effect of reducing any planned repayment of capital.

9. Not only would I have discussed with Mr and Mrs Richards the question of repayment of capital relating to the loan but I would also have discussed with them the position of their current account. I would have regarded the overdraft as providing working capital and no more. It is clear that the overdraft included at an early stage some hardcore borrowing which should have been transferred to a loan account. As mentioned in paragraph 7 I would have insisted on some arrangement to provide for the gradual reduction of the borrowing. In view of the increase in interest rates I might have extended the period for the repayment of capital but I would certainly have agreed some scheme with him.

As a rule of thumb I operated on the basis that a working overdraft should be no more than the annual value of a farmer's milk cheques. In Mr Richards' case this would have been about £40,000.

10. Once the borrowing exceeded £40,000 I would have come to the conclusion that Mr and Mrs Richards were going to experience difficulty in servicing it. It is noteworthy that in November 1979 (by which time Mr and Mrs Richards' borrowing had reached such a level) the Bank advanced £10,000 to Mr and Mrs Richards for improvements to the farm house. I would have advised Mr and Mrs Richards against increasing their borrowing for that purpose at that time because that borrowing would have produced no additional income whatsoever.

11. Once the borrowing was over £40,000 I would have considered that it was appropriate for me to inform my Regional Head Office that Mr and Mrs Richards' account had become a bad or doubtful debt.

12. In my opinion, a Bank Manager owes a duty to a customer to warn that customer if he thinks the borrowing is becoming excessive. It is in the interests of both the bank and the customer to avoid excessive borrowings.

13. On my reading of the Bank's records I have come to the conclusion that the main concern of the branch officials was to cover their own position rather than to advise Mr and Mrs Richards of their excessive borrowing. There was a failure to control the account. The account was allowed to drift. Their attitude may well have been influenced by the fact that because of the increase in land values the Bank appeared to be adequately secured.

In addition to this opinion expressed by a former bank manager, it could be of interest to quote some of the references from Barclays epitome of arrangements and correspondence.

10-6-70 - Farm visited - impression good - steady intelligent progress and attitude.

16-3-71 - A steady shrewd young man who knows his limitations.

15-2-73 - Satisfactory growth from within the business. Has had a brucellosis breakdown.

30-7-73 - Brucellosis problem still rampant; his neighbour is apparently tipping slurry direct into the stream which Mr Richards' cows use.

It is not difficult to understand that many of Watcyn's eventual troubles began with this appalling episode which has already been fully chronicled. For my own part I should hardly imagine that it is something which any decent man would want to have on his conscience even if he were to finish up in life as a chapel elder or M.B.E.

By 11-5-77 it was being recorded that Watcyn was *'in dispute with everybody!!'*

Even so, on 9-10-84 the note said, *'Customer is a good farmer but inclined to be extremely stubborn but he realises the dangers of being overborrowed and every effort will be made to improve his position from now on.'*

Well, of course, Watcyn knew, as he had always known, the dangers of overborrowing, but now that he was asking questions the manager did not like it and there is still no reference to the lost documents. Nor had Watcyn said anything about improving his position because he knew on those figures it was already a lost cause.

Whilst so many of those entries are interesting and significant, two more which have to be mentioned are for August 1986.

12-8-86 – Security on charge form appears to have been altered from 24-12-77 and we do not believe this affects our charge (conveyance date being 24-11-77)

20-8-86 – L.H.O. reply dated 14-8-86. Having spoken to you on the telephone, it is debatable whether our charge has actually been altered, and in any event the local inspection team have never raised a query.

The fact that the local inspection team had never raised a query does not say much for the local inspection team unless, of course, they were covering up in the same way that the manager was covering up for his predecessor, and to suggest that would be to attribute motive.

The important point is that, Barclays having been obliged to produce these documents, Watcyn challenged their authenticity. The legal charge purported to have been signed by Watcyn and his wife. Although neither of them remembered having signed the document, Watcyn accepted that the signatures looked like theirs. It is customary, I believe, in reputable banking circles, when a woman is called upon to sign such a document which could involve the loss of her home, for the signing to be done in the presence of a solicitor who will explain to her the full significance of what she is doing. Most certainly it was not done in this case.

Perhaps more important is the question of the dates, and it seems reasonable to suppose that somebody had changed the date from the 24th December to the 14th because they noticed that Christmas Eve in 1977 fell on a Saturday. Not many banks in this fair and pleasant land, hive of industry though it traditionally be at this glad season, would, I suspect, have been open on that day. And there is even the possibility that the date was changed, not to the 14th, but in fact to the 4th. This point is made because, on March 6th, 1987, the new manager, to whom Watcyn and so many of the Pembrokeshire farming fraternity had taken an instant dislike, signed an affidavit in which he said, on oath, mark you, that the legal charge in question was dated the 4th day of December 1977.

This character, when he had taken over as manager, had made much of the fact that he had taken a special course on agriculture as part of his training and knew all about the subject. It is a pity he did not take a course in chronology as well or instead, because then he might have noticed that December 4th 1977 was a Sunday, and most of the banks I have ever known were no more likely to have been open on a Sunday than they were to have been open on Christmas Eve when it fell on a Saturday. Still, the new manager, an honourable man no doubt, had signed the bit about Affie Davies, as we say in Pembrokeshire, on oath. And I was always brought up to believe that an oath really meant something, even to bank managers and solicitors.

I say this, because when Watcyn's legal eagles were preparing his case for the Court hearing and he drew their attention to these various discrepancies and obvious untruths, they said they were of no material significance, or words to that effect. And who are we to argue? As Robert Service had it to be in The Shooting of Dan McGrew, 'I'm not as wise as the lawyer guys.' And they are the people who are supposed to know about these things. I think.

One further item worth quoting tends to show that the retired bank manager could see exactly what was happening when he had said, *'Their attitude may well have been influenced by the fact that because of the increase in land values the bank appeared to be adequately covered.'*

The reference in this case is to one of the documents, apart from the epitome, which Barclays produced. It was dated 26th June 1980, and the last paragraph reads, *'Form 83 reveals a strong financial base of net worth of £181,000 over liabilities of £45,000, while the borrowings are covered thrice by first class security. The loans will continue to receive regular reductions and it is expected that the overdraft will at least show some reduction at maturity and possibly a substantial one. The borrowings are easily serviceable on overall charge of around £105 per good acre.'*

The out-of-court arrangement having been entered into in March 1987, the date of the adjourned hearing was fixed for July 1st. In June, Barclays asked for, and were granted, an adjournment on the grounds that the Court had set aside insufficient time to hear the case.

The case came before an Assistant Recorder on Sept 15th when the Defendants were granted leave to amend Defence and Counter claim in terms sought, and the application for possession of Bunkers Hill and White Thorn was adjourned to be heard on evidence before a judge.

Also, *'The Defendants do pay the Plaintiffs the current interest of £329.91 per week on a weekly basis and in default of payment the Plaintiffs be at liberty to restore the application for possession.'*

Watcyn's solicitor, having had the benefit of Counsel's advice, wrote to Barclays' solicitors and said they took the view that the condition imposed by the judge as to payment of interest was a wrong exercise of the judge's discretion, particularly since the condition was imposed without the judge hearing submissions on behalf of the Defendants or evidence of their means.

There was more in similar vein, and Barclays' solicitors accepted the argument saying, *'Our client is content to proceed to full trial and we shall thus shortly serve you with our client's list of documents.'* Then the legal wheels kept turning.

As far as Watcyn was concerned the legal wheels turned neither fast nor effectively. A great lethargy and air of casual indifference now seemed to descend upon his own legal representatives. Eventually, after some typical Watcyn prodding, they explained that there was nothing they could do because they were still waiting for confirmation of continuing Legal Aid. As Watcyn was shuttled from one office to another of this outfit he was assured that everything was in order, but still his solicitor received no confirmation in writing.

In the meantime, Watcyn was confronted by one other minor difficulty. Time having elapsed, he still had to carry on with his business and produce accounts in the normal way. Following a telephone call to his new accountants on the subject, however, they wrote to him to say that, as they had been involved as independent accountants in preparing a report for his solicitors, relating to his farming accounts, it would not be right for them to act as his financial advisers as it could prejudice their position. So that, for the time being, he had to seek elsewhere for this service.

The date for the adjourned hearing of the case having been fixed for December 2nd, 1987, Watcyn's solicitor wrote to him to say that Barclays had not yet disclosed the documents in their possession and he did not know whether he would be ready to proceed on that date. In the event, the case was then further adjourned without any reason being given, and the new hearing would have taken place on March 1st, except that 1988 was a leap year, so that it was scheduled for February 29th.

There is no knowing how that might have affected any affidavits being signed on oath that day.

I cannot speak from any personal knowledge of all that happened at that time because, on March 1st, I was far away in hospital in London for a second total hip replacement and had the operation that day. It was, of course, St David's Day, and the prayers of our Patron Saint served me in good stead. It was a pity for Watcyn, though, that it was a leap year because, if his case had been heard on St David's Day, maybe the Patron Saint could have done something to help him as well. The best I had been able to do for him in the way of prayer was to ask my good friends, the Cistercian monks of Caldey, to keep Watcyn and his family in their prayers. With their home and just about everything at stake, they needed all the support they could get, and it will be seen that they did not come out of it too badly in the end. We should never underestimate the power of prayer, and it can be of real comfort and help to those who know that someone is praying for them.

As it was, when it came to court, he was let down badly and, although I cannot write from first-hand knowledge, I have a mountain of papers before me which tell their own grim story.

As the day of the hearing drew inexorably nearer, and Watcyn grew ever more frantic at the lack of action, some faceless character at the Legal Aid department admitted to Watcyn on the telephone that his file had been lost since the previous October. Maybe some of them had served their apprenticeship with Barclays. It would have been the final irony if it had turned up on Watcyn's in-fill area, but perhaps that would be to stretch the imagination too far.

In the meantime Watcyn's barrister, having had a conference with the retired bank manager and examined his statement carefully, offered the opinion that, if the argument was put forward on the duty of care which a bank owed a customer, then it would be a new area of law which had not been fully explored in reported decisions. He therefore recommended that a Leading Counsel should be instructed and that a request should be made for the Legal Aid certificate to be extended accordingly.

Having had clearance at last from the Legal Aid department, following their admission on the loss of Watcyn's file, his barrister said that it would now be impossible to prepare a defence in the short time remaining. Then word came from Watcyn's solicitor that the judge had said that somebody from the Legal Aid department would have to appear to explain the loss of the file. Furthermore, there could be no question of another adjournment because the case had been hanging around for too long already.

Came the day of the hearing and Watcyn was sold down the river. His Q.C. said he had had plenty of time to study the case, thank you, and did not call either the retired bank manager who was present, nor did he call the accountants. And there was no mention of any explanation from the Legal Aid department. No attempt was made to subpoena the two characters from The Onion who had gone with Watcyn to Cardiff on the abortive visit to Barclays Local Head Office.

The Q.C. said Watcyn would win the case if he went

to court, but advised him to settle on the terms being suggested because Barclays would appeal and it would involve Watcyn in astronomical costs which would not be met by Legal Aid. Watcyn has since been assured that costs cannot be awarded against Legal Aid cases.

Watcyn's wife, who had been so loyal and such a tower of strength, wanted to see an end to it all, and bewildered, Watcyn agreed to what was being advised, the terms of which will be considered later, and thereby just about signed away his birthright for a mess of pottage. Having done so, that chapter was closed and there was no way in which the damage could be undone.

Well, there was one way. And that was if fraud could be proved. So Watcyn, the eternal optimist, and still with a naive trust in his fellow-man, went to the Fraud Squad at Haverfordwest. Eventually it was decided at a higher level that no action could be taken, but one of the officers who investigated the case said to Watcyn privately, 'If you'd committed a fraction of these offences you'd have been put inside and the key thrown away.'

Writing in his Way Of The World column in The Daily Telegraph on Nov 21st '92, Auberon Waugh said:

Normally, when I read of large sums of money paid to plaintiffs in libel actions, I am filled with gloom. They seldom deserve so much money – in many cases they deserve none at all – and the survival of the libel laws after Maxwell must be seen as a monument to the cynicism of our lawyers. He used these laws to rob his employees of hundreds of millions of pounds – about 60 'gagging' writs were extant at the time of his death. Yet no suggestion for reforming the libel law has yet been made, more than a year after an object purporting to be Maxwell's carcass was fished out of the Atlantic.

Mr Rupert Allason's victory over the Daily Mirror is a cause for rejoicing. Allason, the young Conservative M.P. for Torbay, was made the subject of a series of violent attacks in the Mirror after he drew attention to Maxwell's links with Mossad, the Israeli secret intelligence and terrorist organisation. After Maxwell had issued a writ against him (it was the last writ the old crook ever issued) Allason felt he had no alternative but to counterclaim.

I remember, some years ago, describing to some French friends an entirely different libel case, brought by a millionaire. When I reached the judge's summing up, they said: "But of course the judge was bribed." It was with some irritation that I explained to them how in England we do not bribe High Court judges. They are pillars of integrity. Even if they weren't, there are the Appeal judges over them – of such integrity as Frenchmen can only dream about; over the Appeal Court are the Law Lords – in the history of England, there has never been the whisper of corruption against a Law Lord. Over the House of Lords, in charge of the entire judiciary, stands the Lord High Chancellor of England, embodying all the integrity that has ever existed in the history of the world.

And when Lord Havers retires as the Lord High Chancellor of England, he takes a lucrative job advising Robert Maxwell on how to issue gagging writs.
Perhaps our self-serving legal system will be reformed one day, but I do not think it is fair to expect the lawyers to do it.

I am grateful to Mr Waugh for his permission to quote this delightful passage.

CHAPTER ELEVEN
A most unusual planning matter

The Inspector who conducted the Public Inquiry concluded the report of his findings by saying, *'In all the circumstances, therefore, I conclude that the Council acted unreasonably in issuing the enforcement notice and that, as a result, the appellant incurred the unnecessary expense of an Inquiry.'*

What Inquiry is that, then, the reader may well ask? Is there really more to which a man has to be subjected? The reasonable or sympathetic reader may not be fully conversant with the bureaucratic propensity of index-linked officialdom to kick a man when he is down.

Reference has previously been made to the fact that Watcyn had eventually obtained planning permission for his new bungalow at White Thorn without having to grease anybody's palm as the saying is. That was in April 1982. And the original intention was that this would be something for the future.

Then, as things developed, there was the thought that the bungalow might fit in quite usefully with the fishing lodges for which Watcyn was still hoping to obtain planning permission. This permission, it will be recalled, was eventually received in September 1988.

It will not surprise anybody with a grain of commonsense that, with all the troubles with which he was having to contend at that time and which have been the subject of these pages, finding the money to start work on a new bungalow was not exactly his first priority. Even so, in the midst of all his troubles with the Nat West documents and the hellish pressure being applied by Barclays in the January of 1987, that is exactly what he did.

This was followed by extremely wet weather, but the footings had been dug out, a trench had been dug and a water pipe laid before work had to cease. Then came the further bank troubles which culminated in his being turned out of Bunkers Hill. And, as the inevitability of this catastrophe loomed sickeningly nearer, the question of finding the wherewithal to build the bungalow did indeed assume a very high degree of priority. How fortunate he was that by this time he was substantially in credit at the T.S.B.

Wherefore and therefore, Barclays having obtained a Possession Order against him at Bunkers Hill, Watcyn deemed it expedient to pull his finger out, as I believe a certain Royal Personage once had it to be, and in March 1988 he recommenced work on the new bungalow. One of the things he had to do as a dutiful, law-abiding and upright citizen, was to send in a card to the Preseli District Council. In response to this he received a telephone call from some character pointing out that planning permission for the bungalow had run out by several months, and please to make a fresh application. This he did and paid £66 for the privilege.

He was, in any case, within his rights or terms of reference or whatever they call it under Planning Regulations, because he had started work the previous year, but he did not argue the point because time was not on his side, and £66 was not going to worry a chap unduly when he was no longer answerable to Barclays whenever he signed a cheque. He did, however, speak to a local Councillor who was a member of the Planning Committee and he said it was all just a formality and Watcyn really only needed an extension to the previous permission and to carry on with the work. Then the Building Inspector called and he said most certainly to carry on building. Rather more to the point, he called in to inspect the work in its various stages as and when Watcyn sent in the various cards, a procedure concomitant with the smooth working of this particular aspect of democratic local government machinery. Five of them there were altogether. Inspection notice cards that is.

Not that there were not other minor considerations and complications. Barclays had given him the heave-ho from Bunkers Hill, so he had to have somewhere to live. His son, by this time had married, but that still left Watcyn with a long-suffering, supportive wife and two slightly bewildered, but typically resilient, young teenage daughters.

The obvious answer seemed to be a mobile home as a temporary measure and, having been assured by the Planning Officer that he would not call down upon his head the wrath of the guardians of the countryside, Watcyn went ahead and set up home alongside the rapidly-growing bungalow.

I must at this point, however, recapitulate, if that is the word I want.

Not having been over enthusiastic at the way things had gone with his court case in the spring of 1988, Watcyn eventually once more changed his solicitors. Before that happened, however, in the summer of that year, and as work progressed, Watcyn decided that instead of using asbestos slates for the roof he would prefer to use Marley tiles. So he visited the Planning Authority's office at Haverfordwest to ask if this would be all right and spoke to the Development Control Officer. I think that is what his title was. Whatever else these characters may lack it is not grandiose titles. And this gentleman told Watcyn there was no question of anything being changed because he did not even have planning permission. It is not difficult to imagine that this pronouncement would almost certainly have resulted in the exchange of some rather harsh words.

The first Watcyn knew about the subsequent developments was when certain gentlemen of the Press came to ask him what was this latest business about the Preseli District Council issuing a demolition order under the terms of which he was required to pull down the bungalow which he had almost finished building. It was also alleged that he was a miscreant of the deepest dye because he had set himself up in a mobile home without permission.

There then followed what may be termed a somewhat frustrating period during which some of the lesser echelons of the local government bureaucracy started to become considerably agitated about something called a Section 52 Agreement. This concerned the fact that the original planning permission had been in respect of a dwelling for use in agriculture. In due season it was established at the Public Inquiry that the Council had been attaching certain conditions to this Section in direct contravention of a directive from the Welsh Office, but this is a small point warranting no more than this passing reference insofar as the Council sought to tie the occupation of the bungalow for all time to those who continued to farm, or be employed, at Bunkers Hill.

Although Barclays had evicted Watcyn from Bunkers Hill he had now managed to rent more land in addition to the twenty acres he owned at White Thorn.

It was in the June of 1988 that Watcyn received a draft of the Section 52 Agreement from the solicitors still acting for him at the time, and it now transpired that a young lady in the Council's legal department was proving to be particularly unhelpful and intransigent, and was demanding to see the title deeds of Bunkers Hill. But these were in the possession of Barclays bank and, in view of all that had gone before, and as recorded in these pages, it will be readily understood that Watcyn had as much hope of persuading Barclays to let him have his deeds as he would have had of selling sand to Saddam Hussein.

Subsequently the same young lady featured on the front pages of the tabloids, as I believe these erudite publications are referred to, as a result of her escapade in doing best part of twenty thousand pounds worth of damage at the home of her seventy-five year old paramour with whom she had had some sort of nocturnal disputation and having at the same time acquired for herself the soubriquet Number One Tart. As a result she was invited to spend six months, without the option, as a guest of Her Majesty at one of the establishments provided for such purposes.

It was, as I have already said, whilst this nonsense was going on about producing the deeds of Bunkers Hill that Watcyn again changed his solicitors. The new outfit leaned over backwards to persuade the Planning Authority to see sense, but all to no avail, and a Public Inquiry was the outcome.

During the course of this completely unnecessary and avoidable charade the barrister representing Watcyn felt constrained to point out that, if the sloppy way the Preseli District Council's case had been presented was any indication of the way they normally conducted their affairs, it was small wonder that the Inquiry had to be called. Amongst other things much play was made of the importance of a plan of Camrose Village. This, however, was not available. A messenger, not a carrier pigeon, was dispatched posthaste to the Council offices but returned with the news that such a plan either did not exist or, if it did, could not be found. There were two photocopies, but these differed from each other in several substantial details.

All in all, it was not perhaps entirely surprising that the Inspector found as he did and awarded Watcyn all his costs.

Maybe it was less surprising that, when the time came to assess these, the legal department of the Preseli District Council, on whose behalf the writer signed with the customary ignorant and indecipherable squiggle, had the gall to suggest that Watcyn had been entirely unjustified in engaging the services of, and calling, expert witnesses. Just as if he would have been content to be guided by the bureaucratic clowns who had been responsible for the whole heap of nonsense in the first place.

Watcyn's solicitors made short shrift of this outrageous contention and the Authority eventually had to fork out £6,733.77.

Let us hope that the poll tax payers of the area felt that they had received, and continue to receive, value for their money.

CHAPTER TWELVE
A shot-gun licence and more pollution

'Without having anything in particular to communicate I take my pen in my hand to tell you what is going on here.'
Thus wrote a gentleman a century-and-a-half ago when writing from the island of Skomer. It was the way they spoke in those days, I suppose.

Such has been the sporadic basis on which this book has had to be written that maybe we were still talking like that way back in those far off days when I first became involved. It certainly seems like a long time ago. It is, for example, a year since I wrote the last chapter.

Having moved into the new bungalow, Watcyn still had to attend to the usual affairs of life on the land. Amongst other things he has for years, like most farmers, held a shotgun licence, but nowadays the law-abiding are all too likely to find themselves being harrassed and penalised because of the doings of the evil element of society against whom the police so often seem to be incapable of acting. There are also some of their numbers, but not all, who seem to see it as their function in life to make the law, as distinct from enforcing it. Bandits and robbers will never find it difficult to obtain and use firearms. I believe that latest reports show that the states in America with the lowest crime rate have no gun licensing laws at all, whilst the highest crime rates are to be found where the gun licensing laws are the most stringent. Not that anybody in his right mind would go much on what they want to do in America.

Be that as it may, during the exercise of moving out of Bunkers Hill and into the new bungalow, Watcyn housed his gun in his son's gun cabinet until such time as he could fit a new cabinet himself.

On April 23rd 1990 he presented himself at Haverfordwest police station and paid the sum of eight pounds in respect of the renewal fee for his licence. He surrendered his old licence and was given a receipt, together with an assurance that a new licence would be forwarded in due course. The cheque was presented for payment five days later.

A few weeks later he was subjected to a visitation from a blue-uniformed female whom he later described in his letter of complaint to the Principal Crown Prosecutor as *'Irma Greese masquerading as a policewoman.'*

On July 18th he received a summons to appear before his betters on the grounds that on May 10th he *'did have in his possession a shotgun without holding a certificate under the Firearms Act 1968.'*

Well, of course, he wouldn't have a licence, would he, when he had surrendered his old one and the constabulary had not yet sent him his new one. Same force, but it is known, I believe, as lack of communication. A common failing, not confined to the police.

Watcyn then went all the way to Carmarthen to the vast new fiefdom of the gendarmerie there to try to sort out the muddle. All he had for his trouble was to be treated by some goon as if, as he said in his same subsequent letter of complaint, he *'had been blood brother to Al Capone.'*

Those who still cling to old-fashioned notions about democracy and British justice could be interested in an important notice which accompanied the summons.

It informed Watcyn that he MUST attend court even if he pleaded 'guilty', and that, if he failed to attend, a warrant for his arrest would be issued.

It was further stated that no prosecution witnesses would be present and, if he pleaded 'not guilty', the case would be adjourned to another date for a full trial.

Fair play, though. The notice did go on to say that it did not in any way affect his right to plead 'guilty' or 'not guilty'. So wasn't that nice of them? And is it not good to know that even the innocent still have some rights left?

The summons was returnable on August 14th at 10a.m.

Watcyn dutifully and punctually presented himself in response to this pressing invitation, but by 10.45 a.m. nothing much seemed to be happening, so he had a word with the usher. That worthy said that Watcyn might not be called until three o'clock in the afternoon, so Watcyn said that in that case he was going home as he had work to do. Then the usher made so bold as to tell him that he couldn't do that and Watcyn pointed out that he had just told him he was going to do it. Then he relented and gave the usher a quarter of an hour to sort something out. So the usher went in to the court-room and spoke to a gentleman who turned out to be a solicitor acting for the prosecution in the case then before the court, and this gentleman came out and asked Watcyn what it was all about and Watcyn told him and the gentleman said 'Good God', or something like that, and for Watcyn to go home and forget about it.

The new date for the hearing was fixed for September 10th, but on August 28th the Crown Prosecutor wrote a very tidy letter to Watcyn in which he said, *'I have received the file of letters from the police which allege that you had in your possession a shotgun without the necessary certifi-*

cate. I have reviewed the evidence and decided that the prosecution will be discontinued and accordingly, I enclose the appropriate Notice.'

I fear that these pages have already contained overmuch of a depressing chronicle on the antics of bureaucracy and officialdom. Suffice it to say that Watcyn submitted a claim to the police for his expenses and the trouble to which he had been put. It was not met in full, but he received a cheque for £75 and, apart from the fact that it was, as we say in Pembrokeshire, better than a smack across the belly with a greasy frying-pan, it proved his point.

Another incident coinciding with this bit of nonsense has been a further disputation about the old polluted watercourse. Yes, the same one that brought Watcyn within the orbit of my ken in the first place. It is almost poetic justice that it should come up again at this late stage.

The last time we heard of it, you may remember, his neighbour had pleaded guilty and got away with a conditional discharge. In the fulness of time he handed over, in theory, to his sons. But, as Watcyn explained when he next had to write to the Water Authority, *'The voice is Jacob's voice, but the hands are the hands of Esau.'* What it meant in effect was that there was no black mark recorded against the sons. And the pollution started all over again.

Eventually, as a result, Watcyn lost four cows. Neighbourly discussion having availed nothing he had to have recourse to other methods.

An appeal to Welsh Water, as anticipated, also achieved nothing, the character whose function it was to attend to such matters being too busy attending meetings. Or so he said when Watcyn telephoned him and, on the last occasion, put the 'phone down. By the time he eventually sent somebody the pollution had ceased.

At the same time Watcyn contacted the Public Health Authority, the dynamic Preseli District Council, who ran true to form in their ineffectiveness. They, too, turned up when the show was over and eventually sent Watcyn a bill for £192.70 for having taken a water sample from a stream in an entirely different parish from that in which the pollution had occurred. One of the idiots there, no doubt fancying himself as a bit of a comedian, had suggested that the pollution had been coming from Watcyn's in-fill area.

Watcyn, however, had remembered the old Pembrokeshire maxim, *'If they haves thee once, shame on them. If they haves thee twice, shame on thee.'* He had, therefore, covered himself by having a report from his veterinary surgeon who had witnessed the pollution and stated that the water was *'extremely murky and discoloured with solid organic matter in large amounts.'*

There was also a confirmatory letter from a land management adviser with A.D.A.S. who said, *'I refer to my visit to your farm on Wednesday, September 9th, and was distressed to observe the extent of the pollution to be seen entering your tributary to the Western Cleddau.'*

Finally, and perhaps even more conclusively, Watcyn had a sample taken by the Public Analysts who issued a report to the effect that the samples were *'badly polluted – due to the very high concentration of organic material present including a high concentration of faecal bacteria.'* Their fee was a mere £40 plus V.A.T., but the more discerning reader will again understand that they would not have the same overheads or carry as much deadwood as such an outfit as Preseli District Council.

When Watcyn was issued with a summons for the

recovery of the £192.70 by the Preseli Authority I suggested that he could do worse than contact somebody like the Flat Earth Society. He finished up with Friends of the Earth and they put him in touch with a solicitor they said was rather good at this sort of work and the case was dropped. He also took on the case of Watcyn's claim against the polluter who was insured by the N.F.U Mutual. Watcyn's insurance broker had warned him at the outset that he had more trouble trying to settle claims with them than with all the other insurance companies with whom he dealt put together. [Speaking for myself, after nearly fifty years dealing with them, my own experience does not confirm that, so maybe it said as much about the insurance broker as about the Mutual.]

Negotiations were somewhat protracted. Watcyn kicked off with a claim for £829, as worked out by an accredited valuer. All liability was denied. Then they said the file had been lost. Then they admitted liability and offered £250. Eventually Watcyn settled for £700 plus all his costs.

Maybe it wouldn't have looked so good in court with the polluter being a bit of a big noise with F.W.A.G., as is the character who approved the siting of the slurry pit in the first place. One slight difference is that the polluter has not yet appeared in the Honours List. These things take time.

Whether matters will improve now that the National Rivers Authority are responsible for controlling cases concerning river pollution remains to be seen, as the monkey said when he left his mark behind the kitchen door.

CHAPTER THIRTEEN
A little white boy in the coal shed

Where now do we turn as we try to tie up the loose ends to bring this narrative to a conclusion?

Since we have just been dealing with rivers and such, perhaps it would be as well to write a bit about Watcyn's fishing lake. You will remember that he started to think about this back in the 1970's. Experience showed that what he had done was not the answer because the lake was too shallow and weeds began to cause the small lake to silt up. Watcyn could see now what was needed and he drew up plans to extend the lake considerably by the simple process of excavating a large area. To his practical countryman's mind nothing seemed to be more obvious or more simple.

Having worked things out and drawn up plans accordingly he then submitted his scheme to the Min. of Ag. and Fish. under the Farm Diversification Plan, and the kind gentlemen at the Welsh Office said yes it was a good idea and no doubt much to be commended by one and all because of the way things were going and Watcyn would receive a grant of £1,193.75, so let us call it £1,200 for the sake of convenience and keep the change.

That was in April of 1989 and there was a stern warning to the effect that the job must be finished by June 30th of the following year which was 1990. Being somewhat enthusiastic by nature and rather quicker off the mark than your civil service types, Watcyn was able to write to them in February 1990 to tell them that he had finished all the work as specified and would they like to call and see it and please could he have his grant. It was only a hole in the ground anyway, so there was nothing to brag about in the fact that the job was done so quickly although it took a bit of time to drive away all the soil that had been excavated.

There were a couple of characters from the Min. of Ag. and Fish. who had seen the work from time to time and they had thought it was pretty good at that. Now, however, Watcyn was told that his application could not be proceeded with for certain reasons and until he had complied with requirements various. He was also called upon to fill in a form, which will not come as any great surprise to anybody, and to put a notice in the London Gazette and one thing and another. He was also required to supply a certificate from a Chartered Engineer registered with the Engineering Council and *'experienced in this type of work'*. This type of work was *'for any raised reservoir'*.

To no avail did Watcyn continue to point out that he did not have a raised reservoir and that had there been any need for a Chartered Engineer he should have been told in the first place before he started and not now when the work was finished.

But, of course, nothing had been said in the first place because they had been dealing with plans for a hole in the ground, and that was what it still was.

It soon became obvious that it was out of the hands of those who had originally been dealing with it, and there was now what we used to describe in the old days, without intending or giving offence to anybody, by the term a 'nigger in the woodpile'. These days such homely phrases are strictly *de trop* as the Frogs say, so maybe we should call him 'the little white boy in the coal shed'. And this little white boy in the coal shed, who had seen nothing of the plan or the work in the first place, became even less knowledgeable and more and more stupid as time went by. Not only did he not understand that Watcyn had driven away only God and he alone knew how many tons of soil from the hole in the ground, but the fool started trying to calculate how much work had been done to build up the sides of the hole, and what volume of water was above ground level, and it soon became apparent that with every letter and every observation he was doing no more than dig a hole for himself.

Before all this happened the F.W.A.G. genius had heard about Watcyn's splendid enterprise, and had turned up to see it and take some pictures, telling Watcyn that he had retired, otherwise Watcyn would not have let him come anywhere near the place. He also sent Watcyn a plan with some suggestions as to what ought to be done to improve matters. In case anyone is interested, this was eight months after he had retired. The report was a load of rubbish, but that was of no consequence. If he wanted to waste his time whilst drawing his index-linked pension that was his own affair. Later, a rather strange thing happened. Presumably in a state of extremis, if that is the correct term, the little white boy in the coal shed called in aid the report from this joker with the suggestion that Watcyn really should have taken note of his report and acted accordingly. It may be considered odd that this could have happened so long after the man had left the employment of the Min. of Ag. and Fish., but the existence of the old pals structure should never be forgotten, and anybody who has read as far as this will not need much convincing that Watcyn was by no means their favourite pin-up.

An impasse having been reached, Watcyn then wrote to the Permanent Secretary at the Welsh Office, and the usual cover-up started, but it meant that the little white boy in the coal shed had to start backtracking at the rate of knots. Whereas his original suggestions would have involved Watcyn in an expenditure of something like £8,000 or £9,000, it now left him with just two stipulations with which to comply. One was that he should remove the gratings and screens from the overflow of the pond as these could become obstructed with debris.

Unfortunately there are authorities in this free and happy society other than the Min. of Ag. and Fish. Watcyn therefore felt constrained to point out that the gratings had been installed to comply with the stringent regulations of the National Rivers Authority. Were this not so the fish could have escaped into the nearby stream. Had these piscatorial escapists been carrying some dread disease they could have infected all the fish in nearby streams and rivers. Possibly a new department could be set up to supervise the immunisation of all captive fishes as a safeguard against such a contingency, but it is not for me to put ideas like this into the heads of officialdom for their further proliferation. And Watcyn opined that, judging by what had been seen of the ability of the little white boy in the coal shed up to that point, it was doubtful whether he could have worked out how to retrieve any fish that escaped. That left the other point, which concerned the *'inlet throttle pipe'*, and we can return to that later.

Throughout these disputatious exchanges there was the strong suggestion that Watcyn should commission a feasibility study. A.D.A.S. would have carried that out for him at a fee of somewhere near the figure of the eventual grant. I am not as close to the heart of farming as I once was, and things are changing at a frightening rate, but I reckon these advisory characters must be in a desperate plight to try to justify their own existence.

In the old days that branch of the Min. of Ag. and Fish was known as the National Agriculture Advisory Service. As a journalist I attended untold conferences, discussions and meetings which they organised. Some of them were good men, and I made, and still have, friends amongst them. But some of them had nothing at all between their ears except daylight. What I deplored above all was their overall policy of calling for more and more efficiency, which meant dispensing with the services of yet another workman, and getting rid of one more enterprise from the farm programme in order to specialise and maximise (a favourite jargon word) the profitability.

One thing they never advocated was that any of their own numbers should be sacked, but eventually, no doubt to their surprise and horror, the idea suddenly caught on at a higher level. The edict went through the land that where farmers now called in advisers from A.D.A.S. their advice and services would have to be paid for. Even assuming that farmers could afford such a luxury, who wants to employ the unemployable? Talk of specialisation has long since been forgotten and the buzz word now is diversification. That is to say that grants may be paid to any farmer willing to opt out of the farmer's legitimate calling of producing food whereby his fellow beings may eat and live. Slow-bellies, and fools who know no better, will no doubt say there we go again with more grants for the feather-bedded farmers.

N.A.A.S. cost much money to keep in being, and that cost was always included in the overall figure which successive governments delighted to quote in order to make people believe that they were doing their duty by those who lived by the land. Only those who did not live by the land were fooled by it.

Along the way they preached the gospel, and the first part of the text from which they preached was that farmers should specialise, as often as not in milk. The fact that it was left to the M.M.B. to market the milk didn't bother the preachers of the gospel, even though it was inevitable to culminate in the introduction of milk quotas.

The second part of the gospel was that, having specialised and increased his efficiency, the farmer could then sack a man. And this continued to be the gospel even when people in other spheres were going through the land preaching the heresy that what was needed was to create employment. So now they must preach from a different text and start explaining to ignorant farmers about wild-life and habitats and so on and so forth. They are also willing to give advice to farmers on the promotion of schemes to make money out of something other than farming, and substantial grants are readily available for such. Yes indeed, more grants for the farmers, and being handed out by people who are experts on any business you care to mention.

Having explained at the outset that Watcyn had come to see me in the first place because of my own experience of brucellosis and the ineptitude of the Min. of Ag. and Fish., it is not perhaps irrelevant to quote something I wrote in Welsh Farm News way back in August 1966. A pilot scheme in connection with the eradication of brucellosis had been put forward, but it had been turned down at a higher level on the grounds that the money was needed to appoint another two hundred and sixty odd people to the N.A.A.S. And the comment in my weekly column was a simple cry from the heart which said, *'Good God above.*

Don't they think we've had enough advice and conferences and demonstrations? Isn't it about time we looked for some practical help instead?'

The cover-up having worked its way down the line from the level of the Permanent Secretary, and victory being in sight at last, Watcyn thought it would be no bad idea to give the Ombudsman's office something to do. He knew, or should have known by now, that he would not receive any satisfaction, but if these people have to be paid it is only right that they should be allowed the pretence of having some sort of justification for their own existence. Their job as far as possible would generally seem to have been established as covering up for their own kind, which is hardly surprising when we remember that their procreator was none other than Little Harold.... *'Can there any good thing come out of Nazareth?'*

Perhaps nice Mr Major will be able to improve on it and sort it all out with an Ombudsman to investigate the Ombudsman.

An approach to Watcyn's local Member for Parliament was therefore necessary. Time was, you may recall, when that office was filled by the then Rt. Hon. Nick Edwards who had reached the dizzy heights as Mrs Thatcher's Nabob in Wales with a seat in the Cabinet, before becoming Lord Crickhowell and being called upon to oversee the disbursement of largesse by the National Rivers Authority, none of which had ever been of much help to Watcyn. The man to follow him was Mr Nicholas Bennett. There are not many Tory M.P's in Wales, so that it was hardly the achievement of the century when he reached the level of Under Secretary at the Welsh Office. And that wasn't much help to Watcyn either, because Mr Bennett seemed to think it was too important an office to jeopardise his position by taking up the cudgels on Watcyn's behalf. He swallowed all the nonsense that the civil servants fed to him and concluded his letter by saying, *'In view of my position as a Minister within the Department responsible for this grant scheme it would not be appropriate for me to take the matter up on your behalf. If you wish to pursue this course, therefore, you will need to approach another Member of Parliament on the matter.'*

This was not particularly helpful, but he ended his letter by sending his best wishes, for whatever they may have been worth, so that was tidy of him. Watcyn acknowledged this reply in August of 1991, noted that Mr Bennett preferred to listen to the lies of civil servants than to protect the interests of a constituent, and went on to say, *'The process of going through the Ombudsman is a lengthy procedure, so that it is fairly certain that by the time that has been concluded you will no longer be M.P. for Pembrokeshire.'*

Thus spake the prophet. At the General Election the following May, the sitting M.P. was informed through the medium of the ballot-box by an ungrateful, if somewhat discerning, electorate, that his services were no longer required.

An approach was then made to the M.P. for the neighbouring constituency of Ceredigion and North Pembrokeshire, Mr Geraint Howells. Initially he said he would take the matter up, but then showed that he was suffering from a dose of the old cold feet disease. He was given the heave-ho as well (and, of course, ordained to higher things by means of the dear old Honours List), at the same time as the luckless Mr Bennett, and was replaced by his Plaid Cymru opponent. And it was a Plaid M.P. who eventually took the matter to the Ombudsman. Mr Bennett having been made redundant (without any mention in the Honours List), the Plaid M.P. suggested to Watcyn that he could now enlist the aid of the new M.P. for Pembrokeshire, Mr Nick Ainger. And why is it, I wonder, that Pembrokeshire should have had three Nicks on the trot as their Parliamentary representatives? Would there be a suitable subject here for a Ph.D. thesis, if it is not a suitable case for the Ombudsman?

Not everybody knows how things are worked in Government Departments. Watcyn was no exception to the rule, and there is no reason why he should have been, but it dawned on him eventually why he was constantly having to deal with two different sets of people. There were the technical characters and there were the executive or administrative wallahs. Once he had established that, he found himself in the position of receiving a letter from one of the administrators saying, as near as damn it is to swearing, that he had to be guided by the technical chap, but it wasn't his fault if the bloke didn't know what he was talking about. After that it was plain sailing. And that all came out when they had come down to the second of the two remaining stipulations, which was the one about the *'inlet throttle pipe'*.

As things stood, and the system was working perfectly, there were two sources of water feeding the lake along an open channel. Some came from a stream which flowed through a wood, and therefore at certain times could bring down considerable quantities of vegetation, which would cause the constant blocking of any pipe inserted to replace the function of the channel. Whereas it was a simple matter to get down into a ditch with a shovel to clear out any blockage from time to time, to clear a pipe would be a different matter. Some of the water was flood water from the road, and was therefore the responsibility of the Highway Authority.

Whatever our individual experiences of these local servants of the public may be, it is gratifying in this case, and it makes a welcome change, to be able to report that they were perfectly reasonable and helpful. That is to say, until the little white boy in the coal shed stuck his oar in again.

And then a compromise was reached. The little white boy in the coal shed had stipulated a six inch inlet throttle pipe, but it was now agreed that a nine inch pipe could be used. Even though he knew what the outcome would be, Watcyn agreed, possibly thinking that once the long-overdue grant had been paid, perhaps the fairies or the little green men would come in the dead of night and dig the pipe up again. Having been assured that a fifteen inch pipe would by no means satisfy the little white boy in the coal shed, Watcyn had a contractor in with a J.C.B. and installed a nine inch pipe. That night in the month of April in the year 1992 there was no more than a moderately heavy fall of rain. The following morning the nine inch pipe was clogged up, but Watcyn had conformed with the bureaucratic requirements, and he wrote to the Min. of Ag. and Fish. to give them the glad tidings and to ask please would they now pay the grant.

Great indeed was the rejoicing when the cheque came, but not many moons had risen clear in the night sky before there was more clogging than somewhat in the region of the new *inlet throttle pipe,* and water, which has an odd habit of finding its own level, was flooding every which way.

Coinciding with this the newly elected M.P. had heard from the Ombudsman to the effect that he could find no evidence of maladministration. What is not generally understood, perhaps, is the fact that bureaucrats are guilty of maladministration only if they fail to answer a letter. It is not maladministration if that which they write to you is nothing more nor less than a load of crap. To use this expression is not in any way seeking to denigrate Mr Ratner's jewellery. As Watcyn explained in a letter of Feb 9th 1990, to one of his multitude of Ministerial correspondents, *'The term "crap", in case you did not know, is in honour of the gentleman by the name of Thomas Crapper who invented the flush lavatory.'*

It did not deter Mr Nick Ainger, the enthusiastic M.P., from trying to do something constructive and, with the festive season of 1992 looming close upon the horizon, in a show of goodwill to one and all, he arranged for a meeting *in situ* between a couple of characters from A.D.A.S., one from the Welsh Office, one from the National River Authority, two from Welsh Water (Dwr Cymru that is), one from the Highway Authority, and I have to stop at that. I read somewhere once, long, long ago, that five is the greatest number which the eye can take in at one time. But I was there as well, because this opus was at last in sight of completion and I had a vested interest as it were.

When introductions had been effected all round, I asked the A.D.A.S. characters why the little white boy from the coal shed couldn't be there, since he was the one who had caused all the trouble. You may find this hard to believe but, I swear by all I hold sacred it is the truth, one of them said, 'He cannot be here because he has been seconded to Ethiopia to supervise a drainage scheme.' And, blasphemous though it may have been, in that moment of complete and utter incredulity, I blurted out, 'Dear God in Heaven! Haven't those poor buggers got enough trouble out there already?'

When one of them used the term embankment, I said, 'Can you tell me, please, why it is that, if an engineer's report had been necessary, you did not tell Mister Richards so in the first place when he submitted his original proposal?' Dead silence. And they certainly could not send to Ethiopia for the answer.

The next ploy on the part of the beleagured A.D.A.S. pair was a technical attempt to blind with science about gradients and levels and one thing and another, but the young gentleman from the N.R.A., who was all there with his lemon drops, pointed out that the figures they were quoting applied only to sewers, and a sewer this was not.

It was the gentleman from the Highway Authority who came to their rescue in the end and said that he would sort it all out in the New Year. He would see to the removal of the offending *inlet throttle pipe,* and there would be no need whatsoever for Mr Richards to call in the fairies or the little green men. R.I.P.

CHAPTER FOURTEEN
Thou shalt not fart in Church

Thinking about Huckleberry Finn, the other day I was saying to a good friend of mine that the horizon still seemed to be a long way away, and he said not to despair because society is becoming more and more concerned over the mounting evidence of chicanery on the part of the banks, corruption in high places, incompetence at all levels from local government up, arrogant and inept judges, miscarriages of justice, dishonest police, overweening bureaucracy, and all the other ills of which we read and hear with depressing frequency, whilst the lip service paid to law-and-order has become no more than a sick joke. So press on with the book, this friend said, because it is the sort of story which needs to be told, and it made me think that he didn't know about poor old Huck either.

Two more episodes only shall I quote, and then call it a day.

Having come to terms with the worst of the slings and arrows of outrageous fortune which Barclays and bureaucracy had been able to hurl at him, Watcyn found it necessary, if he were to continue in business, to produce evidence of his credit worthiness. With depressing regularity, he was informed by all and sundry, and always with the greatest regret and so on and so forth, that by no manner or means could they be persuaded to do business with him. He found this more and more perplexing, because he owed nothing to anybody and was in credit at the bank. On one occasion he was even refused a credit card by The Bank of Credit and Commerce International, so he must have been a pretty bad risk at that. And, of course, nobody would ever give their reasons for their reluctance to trust him, and apparently the law did not require them to do so.

It took time, but eventually he got to the bottom of it. You will recall, no doubt, the interlude of the cattle testing which cost him the sum of £459.50. That debt had been satisfied in March 1987 when Watcyn paid the money into the County Court at Haverfordwest. At the same time he paid two pounds for the record to be wiped from the books. Unfortunately, it remained there for all to see. So Watcyn went in, paid another two pounds and had a proper receipt.

And, in May 1990, having paid another two pounds for the privilege of a search, he received a statement from the Registry of County Court Judgements to the effect that there was *'Nothing Registered'* against his good name.

None of all this, however, would appear to have impressed various possible creditors. They still, with all the usual apologies, evinced a marked reluctance to deal with him or trust him. A reliable associate thereupon took an interest and, in February 1991, wrote to Watcyn to tell him that there was a small matter of £459.50 still registered against his name, from which it may be inferred that, even though he had taken the precaution of demanding a proper receipt for his two pounds, it did not seem to have worked the oracle. A degree of acrimony ensued, and, nothing still having been done, in June 1991 he paid two pounds for another search by the Registry etc. etc., and they came back with the statement that, yes indeed, there was still this small item of £459 against his name.

All he had for his trouble when he went in to the County Court Offices at Haverfordwest to complain was a mouthful of impudence from one of the females there. Some there are who will tell you that they have found out the hard way in this life that rudeness and incompetence invariably go together. Then Watcyn read to her, within the hearing of others who were there at the time, an extract from the Law Society's Gazette of August 1990, which told of the grave disquiet throughout this land of ours concerning examples, which were legion, of the incompetence of the County Courts. One of the sins listed was the failure of court staff to remove notice of payment into court from judge's papers.

One of the Lord Chancellor's recommendations, the report said, was that the Ombudsman's powers should now be extended to enable him to look into the shortcomings of some of these miscreants. The Ombudsman, did he say? Perhaps the Lord Chancellor leads a sheltered life.

The next payment of two pounds in respect of one more search established that the judgement had been satisfied and the record eventually entered by the stalwarts at Haverfordwest County Court in May 1990, and the record shown as having been amended in July 1991. The same missive carried a handwritten note saying, *'Please find enclosed an amended result of search showing the judgement as marked as satisfied. Please accept our apologies for the clerical error in not giving this to you initially.'*

But the Lord Chancellor, apart from his touching faith in the ability of the Ombudsman to sort out some of those who are not up to performing the duties with which they have been entrusted, also decreed that there should now be an *ex gratia* payment to those who have suffered at the hands of some of them, and Watcyn put in his application accordingly. Blessed indeed is he that expecteth nothing, for he shall not be disappointed.

And so we come to the final reference, which is to Barclays. Having obtained possession of his farm, and evicted him and his family from their home, they were still not satisfied, and went far beyond the ruling of the Court by trying to force Watcyn to pay into Barclays the money he was receiving for stock he was selling, instead of into a special and separate fund as ordered by the Court. This, in fact, was what Watcyn had been doing and continued to do. It will be remembered that he still had White Thorn and had also been able to rent extra land. Furthermore they made no attempt to ensure that the farm was sold for the best possible price. There are those, well versed in the law, who will say that there was no obligation for them to do so. Morality would not seem to enter into it. There is no point in going into all the details, but they are available, well documented, if anybody should wish to see them.

In writing of the planning inquiry concerning Watcyn's new bungalow I mentioned that, following the disaster of the County Court case of February 29th, 1988, he had once again changed his solicitors.

Up until the advent of the new firm, Barclays had persistently refused to disclose the price for which they had sold Bunkers Hill. Within a week, his new solicitors, Hugh James Jones and Jenkins of Cardiff, had been informed by Barclays solicitors, Lovell White Durrant, of London, that the price obtained for Bunkers Hill had been £160,000, and they refunded the sum of £12,000. At the date of the Court hearing, Watcyn had still allegedly owed Barclays £135,000.

Over all the chicanery which had led up to the Court hearing nothing could be done. With his wife at her wits' end, and anxious to be free from it all, Watcyn had signed on the dotted line and that was that and all about it.

The Moving Finger writes; and, having writ,
Moves on; nor all thy Piety nor Wit
Shall lure it back to cancel half a Line,
Nor all thy Tears wash out a Word of it.

I didn't write that. It was Omar Khayyam. And we must always give credit where it is due.

There was, however, one not insignificant point where the Cardiff firm could do something, and that was over the charges submitted by the London solicitors, which were considered to be excessive. Lovell White Durrant said they did not agree but, then, as the lady said, 'They would wouldn't they?'

In respect of work undertaken in obtaining possession they claimed a fee of £4,000. Initially they offered to reduce that to £3,000 which, at 25% according to the way we did sums when I was in school, was not too bad for starters.

Lovell White Durrant's second account was in the sum of £4,900 in respect of profit costs, and £40.20 in respect of disbursements.

The business was not disposed of overnight, as will be readily understood. The mills of God may grind slowly, but the mills of solicitors hardly grind much faster. Howsomever, as we say in Pembrokeshire, they ground small enough in the end.

For whatever reason, it was deemed more expedient for the case to come before a Court in London than in Cardiff. Nor was it necessary for Watcyn to attend. In the fulness of time Messrs Hugh James Jones and Jenkins were able to report to their client, Mr Watcyn Richards, that they had appeared before the Court in London on October 17th, 1991, on taxation of both the accounts submitted to Barclays

Bank Plc by Messrs Lovell White Durrant.

In so far as the account in the sum of £4,000 relating to work undertaken to obtain possession following the Order of 29th February, 1988 was concerned, that account was reduced to an amount of £1,200.

The second account in the sum of £4,900 in respect of disbursements was reduced by £1,014.53 including Value Added Tax.

The total reduction was calculated to come to £4,234.53.

Interest was awarded on the amount to be repaid, such interest to be calculated at the rate of 15% from the 7th of November, 1988 till payment. With the interest thus calculated to be £1,868.76 and accruing at the rate of £1.74 per day, it was calculated that the total amount repayable was £6,103.29.

Finally, Messrs Lovell White Durrant were ordered to pay the costs of the proceedings.

All the money could not be released to Mr Watcyn Richards immediately because the case had been conducted with the benefit of a Civil Aid Certificate and the cost of the proceedings had to be assessed.

Watcyn wrote to his solicitors to thank them for their efforts on his behalf, and said, *'It is nice to think that we can win sometimes.'*

When I was a small boy in school, I had an older cousin, very intelligent, very erudite, and very cryptic in many of his utterances. The one that amused me most was when somebody had been given their come-uppance, as the saying is, and he would say, *'That'll teach you to fart in Church.'*

Well, that would amuse a schoolboy, wouldn't it? Funny how I should remember it all these years later.

CHAPTER FIFTEEN
How high do you bounce?

There was a poem we had in school, years and years ago, but I can't remember what it was called or who wrote it. I recall, however, that there was one verse which went,
The harder you fall the higher you bounce,
Be proud of your blackened eye.
It isn't the fact that you're licked that counts,
But how did you fight, and why?
Well, maybe Watcyn has picked up a few black eyes along the way, but I hardly think he's licked. Not yet anyway. It would take better men than most of those we have met in these pages to do that. And I think whoever wrote that particular verse must have had Watcyn in mind as his prototype when he wrote the bit about bouncing back. Or, as the old Pembrokeshire horseman would have said, 'Never let 'em say your mother bred a jibber.'

Above all, far transcending any material considerations, he has come out of it with the marvellous comfort of knowing that he has a wife who has been willing to stand by him throughout the hellish anxiety he has had to endure. In the darkest financial period she was prepared to go out to work as a home-help. Watcyn hated the thought of her having to do it, but it had to be done, and Ray Philippa Richards did it. Together they brought up their three children, who proved to be just as resilient, even when they went through the trauma of losing their home, and the talk about it amongst the other children at school.

Still more important is the fact that, in spite of all life's vicissitudes, the marriage held firm. An inspiration indeed when, in an age of shacking-up, live-in relationships, or 'living comical' as the dwellers of the Welsh valleys far better describe it, almost any reason would seem to be sufficient for forgetting all about the marriage vows. For those, that is, who bother to take the vows in the first place.

Financially, there was never any question of turning to the State. The saying is that those who are willing to work will never want. Until he was turned out by Barclays, Watcyn had the income from his in-fill area. By the time he had finished his new bungalow and he and his family had moved in, he had started to build up more stock and rented some more land. He had also taken on one or two part-time jobs and managed to do the odd deal at buying and selling.

It is not without its shades of irony to record that, having lost his farm, he then received a demand in respect of Capital Gains Tax on the sale of it. How are these things assessed I wonder? I did say earlier, however, in respect of another issue, that the oft-maligned Inland Revenue had been reasonable throughout. And now they ran true to form. They, at any rate, recognised and acknowledged that Bunkers Hill had been sold by Barclays for less than it was worth and the demand in respect of Capital Gains was waived.

Where, then, does that leave Watcyn? It leaves him as the owner, without dispute, of his twenty acres of land, plus a new bungalow, plus a fishing lake with planning consent to build a dozen fishing-lodges thereon. He also has the deeds of same in his own keeping in case any up-and-coming young legal lady should care to see them, and he intends to hang on to them. Furthermore, the entire property is up for sale at a price of somewhere in the region of a quarter of a million pounds.

Once somebody has worked out how to give a kick-start to the economy it will be a whole new ball game on a level playing field, and at the end of the day it will be seen as such in an environmental context. Hopefully there will also be a knee-jerk reaction which will lead to an ongoing dialogue in a grassroots situation and somebody out there will take on board the received wisdom of getting their act together. Or something like that. Hopefully.

When he has sold, Watcyn will then buy another farm without a mortgage, and he will borrow not one penny from any bank on the face of God's good earth ever again. Naturally, in between selling and buying, he will have to deposit his quarter of a million somewhere, and life is full of uncertainties. But one thing I reckon is for sure, and that is that the one bank from which he will not be borrowing is Barclays, and that he will hardly be any more likely to deposit anything with them either.

In the meantime, until the right customer comes along, he could be making a start on the building of the first of the fishing lodges, and he has his eye on a most suitable site for an in-fill area, so we must hope and trust that the banks will treat their discarded documents with a bit of the old tender love and care.

He already has the livestock and some implements, including a Steyr tractor with a safety cab approved by the Health and Safety Executive, as well as by the Ombudsman, so that should be all right. And he will not be milking cows, so there will be no trouble about antibiotics in the milk.

For good measure, nice Mr Major has said he is going to put a stop to some of the sculduggery that has been going on, so maybe in one way and another we can all be looking for a bit of peace and quiet about the place.

You can suit yourselves as to how much notice you take of what I think about it but, for what it may be worth, I reckon that all that sort of talk is so much pie-in-the-sky. The cold facts are very different.

Whether we like it or not, we are faced with the harsh reality that nice Mr Major and his inept bunch of misfits are in so-called power by default, because the only possible alternative is generally considered by the electorate to be too appalling to contemplate.

The totalitarian state has arrived, with the bureaucrats firmly entrenched and in control, and there seems to be nothing that anybody can do about it. Even worse, is the realisation that it has all been allowed to happen during, and in spite of, more than a decade of Tory Government.

Every day we see more and more disturbing cases of people being forced to take the law into their own hands. The police cannot police, and teachers cannot teach, because they spend most of their time having to fill in forms and push bits of paper about. The only real policing is done by the likes of the frightening mob of upstarts known as the 'hygiene police' against whom Christopher Booker continues to wage a worthy campaign in the *Sunday Telegraph*. Worst of all, farmers are harrassed and controlled by a regime that would be a disgrace to a lunatic asylum, and are told to stop producing food, whilst more than half the people in the world are starving.

But never mind. I have done what I set out to do, and come to think of it, even Huckleberry Finn finished his book in the end.

Nunc dimittis ... nil desperandum ... and, in the words of the shop stewards' motto, *nil illegitimae carborundum,* which being roughly translated, means, 'Don't let the bastards grind you down.'